Complexity Avalanche

Overcoming the Threat to Technology Adoption

J.B. Wood

Point B, Inc.

Technology can do even more than it does today to make
the world a better place.
I hope this book helps in some small way.

For P, B, and J.
And to Mamie—you're the best!

Table of Contents

Preface

Ask yourself a question: What is the percentage of all the features of all the technology in the world that are actually being used today?

Got a guess?

Now what if we could increase that number by just 10%? Worker productivity in developed economies around the world would increase dramatically because they could better use their business tools. Undeveloped economies would have a better chance to successfully adopt technology in the first place. Thousands of lives could be saved because doctors and nurses would become more effective at using technology to diagnose and treat disease. Children would learn faster in the classroom and at home through the Internet. The cost of government bureaucracy would decrease at the same time its effectiveness improved. People's income would increase as their technology skills got better. The pace of innovation would accelerate. And you'd finally be able to use your home theatre remote.

The results on the global economy wouldn't be minor; they would be huge.

Now ask yourself another question: Are we getting closer to the goal of people being able to use all of these advanced features, or are we getting further and further away?

Unfortunately, research suggests the latter. Why? It's simple. The world of technology is becoming inherently more complex. This complexity is totally predictable and totally understandable. After all, great technical achievements are rarely simple—especially in a world where instant global communications are taking place across billions of people and devices. It would be nearly impossible to manage all of the variables without some complexity. When asked in a recent survey about changes in the software environment over the last five years, 71% of executives who were in charge of their company's IT systems said it had become more complex. Only 13% thought it was simpler.

Tech companies are great at doing what they love—building innovative products that solve exciting and complex challenges. But what if we reached a point where no one could use the products? Is that where complexity is taking us? And who's working on solving that problem today? The answer is everybody and nobody. Every technology company spends time and money to make their products usable by their customers. But nobody is doing enough—not by a mile.

To be clear, we are not just talking about big companies and their big computers. We are talking about any and every digital product. How often do *you* come in contact with digital technology? Whenever you get in your car, or use the ATM, or visit the doctor, or book a travel reservation, or check your kids' homework online, or look up a phone number on your BlackBerry. Probably hundreds of times a week. What if you were a 10% better user of each of these technologies? You would have more free time in your day, be a better decision maker, know more about what's happening in the world, and have more fun.

The bottom line is this: We need a strategy to make people better end users of technology.

Who should that job fall to? Well, clearly, part of the responsibility falls on the users themselves. They have to take an interest in learning and be willing to pay for some help. Part of it also falls on our educational systems to do a better job of integrating technology skills into their curriculum. And certainly, in the business world, part of it falls on the employer. They need to adequately budget for training not only on technology but on process change—and then give employees adequate time to take advantage of it. But history has shown us that the people who will be the most tenacious and creative at tackling a problem will be the ones who stand to profit most from solving it: In this case, the ones who will profit most are tech companies.

You see, if users can't get the value from a product, the product will fail in the marketplace. On the other hand, the more value they do get, the more successful that product will likely be. It is in the best interest of tech companies and their shareholders to care more about—and invest more into—user success. Unto itself, this can be a profitable activity. But perhaps even more importantly, it is central to the continued vitality of every sector of the technology industry. And with digital technology creeping into virtually every product category, this becomes a mighty big issue for everyone.

Every person that I have talked to—especially those who are not in the tech business—has had the same reaction to the idea for this book. When they first heard the name, there was a look of bewilderment. Then, in a few seconds, a different look appeared on their face. A moment of realization hit them. "You are exactly right," they said. They could all go on to name product after product that they can't use well or get to work at all or get to work together.

At some point, both consumer and enterprise customers say "Enough is enough!" Do they alert anyone? Maybe. Do they stage a protest or ban together with other users in an organized revolt? No. They just stop spending more money on that product and maybe tell a few friends or business peers about their experience.

The first thing we need to do is to acknowledge that we have a growing problem with complexity. Then we need to work together to solve it. If we do, everybody wins—customers, companies, employees, shareholders, and society. If we don't, we will undercut the potential good that technology can do. There is no need to let that happen.

Acknowledgments

I WANT TO OFFER SPECIAL THANKS TO THE PEOPLE AND THE COMPANIES who helped us put together this book. First, thanks to the whole team at TSIA and the more than 30 reviewers of this book for their contributions. A few special callouts:

- Chris Dowse, CEO, Neochange—a true thought leader on these concepts.
- Thomas Lah, Executive Director, TSIA—for many great ideas and great refinements.
- Ron Ricci, Vice President, Corporate Positioning, Cisco Systems—for giving me the courage to say more.
- Bill Steenburgh, Senior Vice President, Xerox—for being the leader he is.
- Jim Spohrer, Director, Global University Programs, IBM—for pulling the academic world together on the subject of service.
- Geoff Moore and Todd Hewlin, Partners, TCG Advisors—for being smart, helpful guys.
- Armin Brott—for getting me off on the right foot.
- To the 400 member companies of TSIA—the thoughts came from all of you.
- To the makers of wine—it takes a lot of great wine to make a good book.

1 | The Consumption Gap

It turns out that technology does have its limits. Not because engineers can't innovate, but because users can't use. And it is costing tech industries billions in revenue growth every year. The gap between the value that technology products have the potential to deliver and what customers can actually achieve is growing rapidly. Most customers are struggling to keep up, and they usually settle for far less value than they could (and should) get from their purchases.

Unfortunately, most tech companies today lack an effective plan for driving customer success. Why? It's partly because they can't get clear of their own product DNA. But it's mainly because of the organizational constraints imposed on them by their current financial models. Their business strategy simply won't allow them to do what's really best for the customer. Sure, they are great product innovators. But delivering true success to customers today requires much more than cool technology—and that is where the breakdown occurs.

A new business model for the tech industry is unfolding—one that requires radically different thinking about the future of services, sales, R&D priorities, and how companies create shareholder value. One that views the use of the product as the beginning of a journey with a customer, not the end. One that defines success in the customer's terms, not based on revenue recognition rules and customer satisfaction surveys. One that creates competitive differentiation and profits not by adding more features but by getting better results for customers from the features they already have.

The companies that effectively help their customers close this value consumption gap will be the next winners. Feature-based differentiation is fading. Results-based differentiation is rising. Fortunately, many of the pieces needed to deliver this new model profitably are already in place and paid for.

But first, we need to take a step back.

Over the past two decades, the world has seen the digitization of nearly everything. There are the obvious things like computers, software, cell phones, and iPods. But today, cars are digital. So are toys, medical equipment, manufacturing lines, multi-function copiers, TVs, aircraft controls, and musical instruments. And innovations like GPS have made the most low-tech things of all—like taking a hike in the woods—a digital experience.

From the manufacturer's perspective, the shift to digital is great news. Once a product goes digital, companies can add new and amazing features faster and cheaper than in practically any other form of product development. No factories to build, no dies to cast, no natural resources to deplete. Nearly every industry already has (or soon will find) a way to create a digital component to their product. Maybe it's in the product itself or maybe it's in the way you manage the product—like ordering office supplies on a Web site. Beginning on that day, you can count on a rapid proliferation in the features and capabilities of that product. First it is just some basic features. Soon new features will be built on top of the last ones, and so on and so on.

Don Norman, author of *The Design of Everyday Things*, says there are three primary reasons why companies focus so much of their resources on adding new features into their products.[1]

1. **Customers ask for them.** "If the product could just do x, y, or z, we'd buy it." That also means that for every vertical market, or every horizontal market, we add features. The more markets we want to service, the more features we add.

2. **To trump competitors who are also adding features.** This is a phenomenon as old as business itself: The product with the most innovative features at the lowest price wins.

3. **Engineers want to show they can do it.** Product developers and teams have their own sense of pride and accomplishment. That usually takes the form of technical achievements, many of which have created fortunes and notoriety for the engineers and the companies they work for.

From the customer's perspective, the all-digital world is a mixed blessing. On one hand, adding feature after feature has made products more capable (and in the case of consumer products, cheaper and more fun). On the other hand, all those new features are creating an avalanche of complexity that's growing bigger and moving faster in industry after industry. And it doesn't stop with individual products. In fact, one might argue that there's almost no such thing anymore as an individual product—they're all just components in larger networks. And the benefit of that network is what customers really want. That, of course, makes things even worse. If you're trying to get three already complicated products to work together, instead of making your life easier, you've got complexity cubed. In order to be successful, customers from global corporations down to consumers need a rapidly escalating degree of expertise and experience. That is true because this avalanche of complexity can be seen everywhere: from complex corporate IT networks to home theater systems to cell phones to medical imaging instruments to, well, you name it.

It all begs the question: How are customers—whether they're enterprise, small business or consumer, CIO, doctor, shop supervisor or student—surviving the complexity avalanche? According to a number of recent studies, not particularly well. Consider these stats:

- According to a 2009 survey conducted by TSIA, Neochange, and the Sand Hill Group, only 14% of enterprise software deployments are rated as "very successful" by the company IT executives who own them. Of these customers, 12% rate themselves as "not very successful" and 74% as only "moderately successful."[2]

- Research group NPD reports that while 71% of mobile phones sold in the United States in 2008 had video capability, only 28% of users were aware they had that feature.[3] Awareness among consumers that they can connect their phone to a local Wi-Fi network is probably similarly discouraging.

- According to a survey conducted by British Telecom in 2008, 71% of Britons have up to 10 gadgets lying idly around the home, as they find them too hard to use. The study also shows that 94% of people who experience problems with their home IT are too intimidated or proud to seek expert help. Over 80% of those who have a problem try to fix it themselves, or ask family and friends for advice.[4]

- Only 5% of consumer electronics products returned to retailers are malfunctioning—yet many people who return working products think they are broken. The report by technology consulting and outsourcing firm Accenture pegs the costs of consumer electronics returns in 2007 at $13.8 billion in the United States alone, with return rates ranging from 11% to 20%, depending on the type of product.[5] It's not so much that the product itself didn't work; it's that the customer couldn't figure it out—this is especially true when it is part of the complex network we all know as "home theater." Here is how consumer returns due to "no fault found" have trended over the last 25 years.[6]

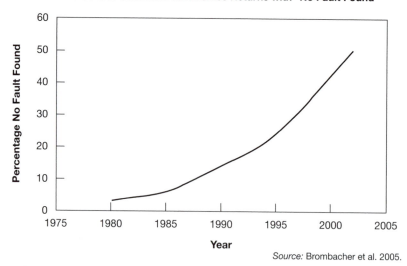

Source: Brombacher et al. 2005.

FIGURE 1.1 Percentage "no fault found" in modern high-volume consumer electronics.

- Just a few years ago, the BMW 7 series landed up with the dubious distinction of making *Time* magazine's 50 Worst Cars of All Time list. Here is what Time.com said: "Perfectly constructed, astonishingly fast and utterly besotted with technology, the big, gracious 7-series had … flaws: The first was something called iDrive, a rotary dial/joystick controller situated on the center console (based on the Windows CE operating system), through which drivers adjusted dozens of vehicle settings, from climate, navigation and audio functions to things like the sound of the door chime. The reason for iDrive and similar systems is that designers were running out of room for switches and instruments. The trouble was that the iDrive was hard to work. Damn near impossible, in fact. Drivers spent many hair-pulling minutes driving to figure out how to add radio presets, for example, or turn up the air conditioning. When confronted with complaints, BMW engineers said, with barely disguised contempt: *Ze system werks pervectly. Dis is no problem.* Since 2002, BMW has gradually improved iDrive to make it more intuitive, but it's still a pain."[7]

- In another enterprise case, a major software company is reported to have less than one-third of its sold licenses in actual use. In other words, more than two-thirds of what they sold is not being used. Again, it is not because the products don't work or they are not useful, they simply are not being adopted. In such a case, how many incremental purchases can the company expect from their existing customer base over the next few years? Not many.

What other industries besides tech could possibly survive—let alone thrive—with customer results like these? Could Boeing stay in business if just 14% of the airlines were "very successful" at getting pilots and crews to use the features of the plane? What about John Deere with its tractor customers? What about your company? Would you tolerate only 14% of your customers being "very successful"? Probably not.

In their defense, tech manufacturers and software companies—both business-to-business (B2B) and business-to-consumer (B2C)—have tried creating voluntary technical standards to improve user interfaces and interoperability. Over the years (and years and years) standards do tend to happen and are hugely beneficial. But this long delay means that standards really haven't done much to hold back the complexity avalanche. Standards simply can't keep pace with the rapid proliferation of great ideas from smart companies and the features and capabilities that flow from them. For the first time, some tech customers are actually interested in having features taken out!

You will notice throughout this book that we bounce between product categories, industries, customer types, and price points. That is because the complexity avalanche is not just happening here or there; virtually *every* technology category is in its path. Many people would argue that this problem was once the domain of big enterprise IT customers. Now it's happening in the home. And if it hasn't already, it will soon happen to you.

THE CONSUMPTION GAP

So here's the dilemma facing technology companies large and small: If your end customers can't figure out how to use your product or they can't get it to work in their network or they can't change their business process to adapt to its features, it has little or no value to them. No amount of slick graphics, flashing lights, and jaw-dropping technological advances is going to change that. The ultimate goal for technology companies is no longer just to sell products. In order to make customers successful, companies must move beyond that thinking. The ultimate goal today is to enable customers and their businesses to get full value out of the product, the value it has the potential to provide.

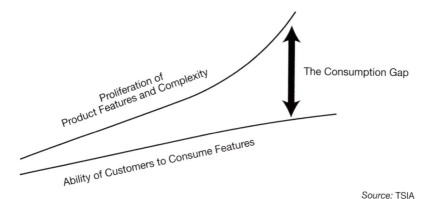

Source: TSIA

FIGURE 1.2 The Growing Technology Consumption Gap.

The difference between the value the product *could* provide to the customer and the value it actually *does* provide is what we call the "consumption gap." And the consumption gap is growing, not shrinking. We're talking about the manufacturing company who has only half its plants using its scheduling application, the IT guy who can't use important system utility functions in the network management tools, the MRI tech who takes twice as long as his peers to do complicated scans, the average digital camera user who has only a small percentage of digital images in print, or the fire captain who can't use the central dispatch system in the fire truck at all. Solving problems like these is a challenge that should rally your company no matter what its role in the supply chain. Whether your company makes tech products, relies on tech products, or sells them direct to customers, the consumption gap is a growing threat to your business.

Here are five powerful reasons why you and your company might care deeply about the consumption gap facing your customers:

1. It might be limiting the size of your market to only the technically advanced, "early adopters." They are the only ones willing and able to figure things out.

2. It might be slowing the rate of repurchase by your existing customers because they aren't using what they already own.

3. It might be increasing your cost of sales and reducing your product margins on repeat purchases because the value of their old version of the product wasn't fully realized.

4. You might be leaving service revenue and profits on the table and missing a huge differentiation opportunity. Customers want help and will pay to get it.

5. If your sexy new features aren't being used, they aren't much help against the competition. There are multiple examples of products that compete effectively with other products boasting twice as many features. Both have the few features that customers actually use. The rest are not just useless, they are actually negative to some customers because they add clutter to the interface and steal system capacity. And they cost money to develop and maintain, used or not.

We are going to talk about these realities (and others) in different industries throughout this book because we see it happening in market after market. Once a product goes digital it is only a matter of time before a consumption gap builds up around it. The more digital features in the user interface, the larger the likely consumption gap. And left unchecked, this consumption gap begins to eat away at your company's effectiveness. In the extreme, the consumption gap could actually kill a product.

The easiest place to see this in action is the lucrative world of enterprise technology. That's because enterprise technologies usually offer two value propositions: one to the end user and one to management. As an example, let's take CRM software.

A CRM tool gives salespeople the tools to organize their tasks and stay current with customers. That same tool also provides the company's management the insights they need into the sales pipeline so they can better predict the next quarter's sales. If the end users—in this case, the salespeople—opt out of using the tool, management doesn't get its benefit. To be successful—to get full value out of a million-dollar software purchase—corporations need to get end users to use not only the features they like but also many that they don't. If that goal isn't achieved, large parts of the product's value proposition go unrealized. Gartner, a leading technology research and advisory firm, predicts that "through 2010, 75 percent of CRM SaaS [software as a service] deployments will fail to meet enterprise expectations."[8]

In fact, one S&P 500 company we know of spent over $20,000,000 to purchase and implement a CRM solution, then had a mutiny in its sales force once it went live. It ended up writing off the whole amount in a restructuring charge and went with a simpler solution that the salespeople could actually use.

But it's not limited to large companies. In the case of our fire captain, his inability to use the technology might make him unaware of the status of additional equipment that is en route, it is adding to dangerously overcrowded radio traffic because on-scene personnel are verbally relaying commands to central dispatch rather than using the system, and it is almost certainly screwing up a lot of activity reporting at the district level.

Even the home today practically needs a full-time IT staff to answer user questions, install upgrades, maintain virus protection, get the TiVo to record only the new episodes in a series, and get the home theater speakers to work with your iPod.

FEATURES ABOVE THE FOG LINE

In Silicon Valley, it can get foggy. On those days you don't see what's happening above the fog line. There are all kinds of cool things going on up in that blue sky beyond the low clouds. A beautiful bird might swoop majestically around in circles; you'd miss it. An enormous plane might be leaving a silvery orange jet stream from the eastern horizon to the western one; you'd miss it. Superman could fly by; no one would notice. When it's foggy in Silicon Valley you pay attention to what's happening below the fog line. You are blind to the rest.

It turns out there is a different fog growing over Silicon Valley and other kinds of technology markets around the world. It's not one that the sun will burn off, either. It's being driven by complexity—technical complexity, process complexity, and features complexity—that is quite simply overloading customers. These complexities are now creating a fog that is hiding the features of technology products in consumer markets, business markets, medical markets—anywhere you find high-featured, digital products in end-customer use. More importantly, they are preventing the benefits the customers should be receiving. What exactly do we mean by these complexities?

- **Technical complexity**—what it takes to install, set up, integrate, migrate data to, maintain the performance of, and keep up-to date a digital product or network.

- **Process complexity**—how companies have to change the way they do parts (or all!) of their "quote-to-cash" business processes to make them work with a particular technology. Or the way consumers have to organize and store their digital photos. Or the way doctors have to change how they access a patient's medical history in an examination room.

- **Features complexity**—the difficulty of learning to use the features themselves. Like trying to learn a cool new way to sort a customer's transaction history using the "simple" 17-step, nine-page section of the user manual. Or loading your old data onto your new system. Or ... well, you get the idea.

In each of these aspects of complexity there is a growing consumption gap between what digital products require their users to do to achieve full value versus what they are actually able to do. This is true from budget-constrained enterprise IT organizations to time-constrained professionals like doctors and engineers to experience-constrained consumers who like photography or want to do their own taxes. Just look at these results from a research project commissioned by Cisco to study what the keys are for enterprises to increase productivity gains with IT investments.[9]

Start With...

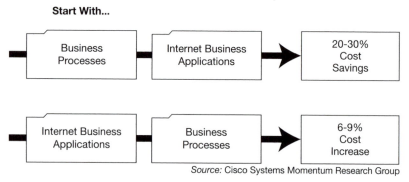

Source: Cisco Systems Momentum Research Group

FIGURE 1.3 Q: What Is Most Fundamental to Achieving Productivity Gains with IT Investments?

They found that the order in which you took action on a project—change the internal business processes first and then implement the applications versus implement the applications and THEN change the internal business processes—had a huge impact on the cost effectiveness of the total project. For those companies who changed the business processes first, the total project resulted in an average cost savings of 20% to 30%. But those who implemented the technology first actually had their costs INCREASE an average of 6% to 9%. Clearly, increasing costs was not the goal of the project. We would argue that this is a classic case of the costs of the consumption gap. By forcing the organization to change its processes first, the changes actually got made and adding the applications was easy. Large parts of the consumption gap were eliminated even before the technology showed up. But go the other way around and the pressure on the process changes often get lost. Because the "implementation" is complete, the whole project can be depressurized. The tech vendor pulls out and the customer is left working "around the application" instead of through it. They are actually adding processes and using alternatives to the application to survive because they haven't fully adapted the organization and gotten end users to adopt the features. The consumption gap is robbing them of their cost saving but there is not that much emphasis on or investment in closing it. This is why Lean Six Sigma works so well. It forces companies to "lean out" the process first—to eliminate the non-value-add steps, thus making it easier to change.

So where does that leave these enterprises? Well Cisco went on to ask what the greatest barriers were to future productivity growth.

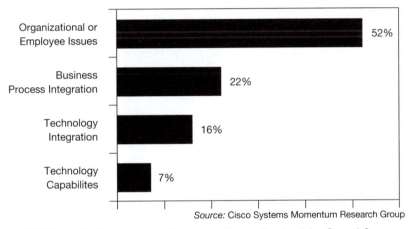

Source: Cisco Systems Momentum Research Group

FIGURE 1.4 Q: What Are the Barriers to Future Productivity Growth?

Guess what? It was not that the technology lacked the capabilities, or that it didn't integrate, it was about processes, culture and people—basically, barriers to end-user adoption. Now you would be right to insist that there is inherent complexity in any large enterprise's business processes. They are large, complex organizations and it is damned hard to get them to change. You could also go on to say that there is a great deal of inherent complexity inside an enterprise tech company as well—many products, many markets, many channels. These all add further challenges to solving the consumption gap. But, as we will point out in Chapter 4, there is a certain absurdity to the difference between where technology customers face their biggest challenges to success today and where technology companies are investing their money to provide services to those same customers. We believe that many are simply not working on the right problems.

We believe that to millions of business tech customers—and consumers—around the world, closing the consumption gap will become one of the most important considerations in their selection of digital products and vendors. Who cares about supposedly great features if they are not being used? The psychology of markets is shifting. This is not just true in a few markets but in all manner of technology product categories where the customers are moving from "gadget buyers" to realists. They are becoming more worried about getting the help they need to be successful than which product has which feature.

At the Technology Services Industry Association (TSIA) we deal with leading technology companies every day, and it's extremely rare that we find one that doesn't have a growing consumption gap. What about your company? Here's an exercise you can do in the course of an afternoon that will give you some eye-opening insights into the extent of your consumption gap.

Go find a couple of your product managers and ask them these two simple questions about the product they are responsible for:

1. What is the combination of our product's features that—if all our customers used them—would give us maximum competitive advantage, customer satisfaction levels, and product and service margins?

2. What percentage of our customers actually uses all those features?

Happy with the answer? Probably not. Now take a walk over to one or two of your sales executives, and ask them to tell you what is happening to sales cycles and the cost of sales. If you're like many companies, this news won't be good either. They'll tell you that customers are now talking more and more about getting value out of the products they already own, rather than about adding newer models with even more advanced features (that they won't use).

If you want to dig a little deeper, go out and talk to your customers. You'll probably hear many of the same things that the mobile device industry did in 2008 when they read the results of the industry's largest research project, "The Global Mindset Survey," which interviewed 14,000 users in 37 countries. It turns out that new features are just about the *last* thing on customers' minds. "Frustrated with complex devices overloaded with hard-to-use features, users said they would find a way to make them simpler and easier to use," writes the study's author.[10]

Take even the basic example of the remote control in your living room. If it's too hard to use, would you or your spouse recommend purchasing another one from the same company? Chances are you aren't too impressed. If you're the manufacturer, it's tough to change that impression about your brand. If enough people reach the same conclusion, your product has a problem in the marketplace.

Once again, it's important to remember that the consumption gap is *not* limited to traditional technology markets like computers, networks, and software. The 2009 Lexus LS sedan now has an owner's manual that comes in at a whopping 700 pages. As a result of this increased complexity, Lexus has had to offer the following new service to customers:

> *Lexus Personalized Settings: Lexus vehicles include a variety of electronic features that can be personalized to your preferences. For example, doors can be programmed to remain locked when you shift into "Park." Programming of these features is performed once at no charge… Programming of your Lexus Personalized Settings requires special equipment and may be performed only by an authorized Lexus dealership.*

Some Lexus dealers around the country have even begun dispatching service reps to the customer's home in order to perform these services, which are essential to owner success and which they can't or don't do on their own. So whether your business is planes or trains, medical devices or telephones, and whether you manufacture products or sell them, you had better take stock of your consumption gap.

THE CUSTOMER AS GENERAL CONTRACTOR

Much like getting a new home built, getting a technology solution today to deliver its full value potential can be complicated. It involves a large number of steps that require time, skill, and resources. In the home construction industry, you hire a general contractor to oversee that complexity. The contractor interprets the building plans, hires and schedules expert subcontractors, ensures that appropriate processes are implemented, provides quality- and cost-control functions, and delivers a usable end product: a home you can live in.

When it comes to getting digital technologies up and running and providing full value, who acts as the general contractor? Usually it's the customer. In our recent survey, 77% of corporate IT buyers reported that indeed *they* were primarily responsible for getting value out of their technology purchases.[11] Customers—who presumably bought the product in an attempt to save time and money—are now forced to find the time, locate the IT experts, manage the project, watch the costs, change the processes, and get it all right.

Forcing customers to be their own general contractor is by no means confined to large IT deployments. What about consumers who want to learn how to use a Web site or a new mobile device? They are in the exact same predicament.

Some customers have banded together in user communities to help each other overcome the complexity avalanche. They ask questions of the group, "Google" the problem, and hope that other customers who have had the same problems might come to their aid. That may work in simple cases but, far too often, the customer-as-general-contractor approach fails.

A THREAT TO PRODUCT-BASED DIFFERENTIATION?

Is the complexity avalanche avoidable? Probably not. Complexity comes naturally once a product category begins to go digital. But the impact of the consumption gap is threatening to upset a lot of dynamics in the tech business. As an example, the age-old strategy of trying to grow market size and share by adding features and functions is now being inhibited (even gated!) by the growing consumption gap. As tech companies, we are in danger of putting our most important new capabilities out of reach of our customers. Would a grocery store ever put its hot selling products on the highest shelves where customers couldn't reach them? Absolutely not. It puts them right at eye level so you can't miss them. Or it will stack them up near the checkout aisle so you practically trip over them on your way out. It makes them simple to consume. But tech companies haven't learned that lesson. Many of them are guilty of putting their coolest and most innovative features out of the customer's technical reach—they're too hard to learn, too hard to connect, and have too many dependencies for their typical "main street" customer. You may already have product features that differentiate you from your competition and are important drivers of customer satisfaction and purchase behavior but which your customers aren't adopting. Or maybe you have heard your customers request product "enhancements" for features already built into the product.

As we said, a growing number of customers—especially in the enterprise market—are starting to move from "dazzle me with features" to "show me how to get the most out of what I bought." Industry after industry is reaching a tipping point where users and customers are changing their preferences, their buying habits, and their psychology. They're paying close attention to how their purchases get used, and they're selecting vendors based on how successful they are at maximizing value to the customer. So if you plan to dominate the competition through features superiority, your company needs a strategy to bridge the consumption gap and increase the value your customers ACTU-ALLY get when they buy something from you. How do you do that? By doing something that *sounds* simple: Start selling results through new kinds of services that you don't offer today.

A PROBLEM IN THE VIRTUOUS CYCLE

As tech companies, we want our customers to engage in a very simple pattern of behavior, which we at TSIA call the Virtuous Cycle. The Virtuous Cycle calls on the customer to follow these four steps:

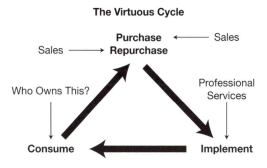

FIGURE 1.5 Services Will Plug the Hole in the Virtuous Cycle.

Step 1: Purchase

Step 2: Implement

Step 3: Consume

Step 4: Repurchase

Most healthy businesses are successful at getting their customers to adopt the Virtuous Cycle to one degree or another. We invest extensively in organizations designed to help customers through this process. In Steps 1 and 4 we spend millions (or more!) to provide customers around the world a sales representative, team, or dealer network to help them successfully navigate the internal and external challenges of procuring our product. Similarly, most companies have at least one service organization that's available to the customer during the implementation phase. In enterprise, commercial, and government markets we or our partners have large, global professional services (PS) organizations to help our customers achieve their desired result: a usable system. In the consumer space it is a 24/7 800-number or a Web site. All these investments are great and worthwhile from a customer success perspective.

But ask yourself an honest question: What organization in your company is responsible for driving your customers' successful consumption of product value? It isn't sales: Their attentions are often drawn off to the next deal once this one is closed, which is probably just what you want them to do. It isn't the professional services team; they're so driven by being billable that once the hours in the project have been used up, unless they can sell another engagement to that same customer, they're off to the next billable customer almost regardless of whether the last one is actually deployed or not. And don't fool yourself into thinking that it's your current customer service organization. The objective you gave this group is to maximize product availability and performance to customers by meeting your service level agreements at the lowest possible cost to your company. The prevailing view among tech CXOs is that customer service is a cost center and should be managed as such. Unfortunately, that particular goal often leads customer service managers to develop tactics to avoid or "deflect" calls, to minimize talk times, and to offload consumers' questions to online user communities. In short, to *not* deliver service. So even in the case where senior account executives (salespeople) of a large enterprise tech company do take their customers' success very seriously, what can they really do beside yell loudly on their behalf? What internal organization could they go to today whose business purpose is to drive more success for their customers?

But what if things were different? Imagine what would happen if technology companies had an organization that was chartered to accelerate the consumption of product value to help achieve customer results—not to avoid supporting the customer but to seek the opportunity to make the customer more successful. We would be incenting our service reps to proactively seek out customers who need help but haven't yet asked for it. We would reward them for making our product truly valuable to our customers. After all, isn't that what you really want? It is certainly what your customers really want. And yet in most companies, there is no organization that is tasked with delivering this critical outcome.

At a growing number of important companies, executives are seeing that the key to winning future share in their markets will be determined not by products and features, but by the services that are wrapped around them.

Unfortunately, most high-tech manufacturers and software companies aren't ready to take this critical step. They cling to the idea that differentiated product is all that truly matters. Others think they already provide services to address problems like these. We at TSIA respectfully challenge that assumption. You may provide a long list of services, but we're betting that those services are aimed at a different objective. That means that your company's inability to help your customers navigate around the complexity avalanche will become a gating factor to your growth—maybe even to your survival. The business model for tech companies is shifting. What we do and how we make money is about to change. Because just selling products is a dying strategy.

2 | The Money Moves from Products to Services

THE CONSUMPTION GAP IS JUST ONE OF MANY FACTORS THAT IS CHANGING the way today's large tech companies—or pretty much any organization that sells anything digital—run their businesses. Success in rapidly maturing tech markets, especially in a difficult (and some say fundamentally changed) economy will require tech executives to do some serious rethinking. For example, many companies have already begun to realize that:

- Existing customers have become the priority customers.

- What their customers need from them—customers' definitions of value, success, and supplier qualifications—is evolving in a new direction.

- Technology vendors must move from transaction-oriented thinking to ongoing services-oriented thinking—not just in how they charge for access to their product but what they actually do for the customer. Their revenues and margins will also follow that transition path.

- The fastest and best way to win market share, improve product margins, and reduce the soaring cost of sales may be to invest more in their services business, not relentlessly cut its budget. Services can deliver better total shareholder returns if they are chartered to profitably grow total revenue (products and services), rather than just be managed as a cost center.

As we all know, success often requires that you be willing to adapt and change yourself and your organization. In the era of the complexity avalanche you may need to reevaluate the way you think about your company's offerings, the way you spend your R&D dollars, and even the way you measure the health of your customer base. (More on this important last point in Chapter 7.)

It is easy for companies to convince themselves that they have already embraced the concepts in this book, that they are old news. We would contend that you can count on one hand the number of tech companies that have truly done so. And in each case they are stealing market share. IBM is one. HP bought EDS to catch up. Apple with its Genius Bar is another. Now Sony, Microsoft, and others are following the model. Even Xerox is fundamentally re-thinking their strategy. The point is, not that many global brands have their money following their public statements. But the fact that it is finally happening is compelling evidence that the rise of services is real and that companies need to make aggressive moves to remain competitive. Just in case buying EDS is not in your budget, we have a few other ideas about how you can use services to gain market share for your products.

CHANGING YOUR PRIORITY CUSTOMER

In the early stages of a new technology's existence, it is all about new customers—without a steady supply, your company would have gone out of business. Customers gave you the revenue, credibility, buzz, and market insights that put you on the map. It makes perfect sense, then, for emerging companies—or product lines—to say that almost 100% of their focus should be on new customers.

Then, voilà, some of those new customers came back and bought from you again. Soon the percentage of revenues from new customers dropped to, say, 90%, while the percentage from existing or return-ing customers increased to 10%. Over time, that ratio got smaller and smaller, and eventually you got to that wonderful point where exist-ing customers became bigger contributors to your company's top and bottom lines than new customers. This happens to every successful company. And if your company or product line is just emerging now, it will eventually happen to you, too.

The good news is that a lot of the revenue generated by existing customers comes from higher-margin items, such as upgrades, supplies, replacements, and maintenance contracts. In addition, these repeat orders usually come with a much lower cost of sales. Over time, this means that a company's profits should rise, even if revenue stays flat, thanks to a more profitable mix of existing versus new customer sales. In reality, though, it's more likely that revenues will *increase* as the new-to-existing customer ratio shifts toward existing customers. Loyal customers—those who know your brand and love what you're selling—tend to buy from you more frequently, are often less price sensitive, and tell their friends. While price pressure is at a premium in this tough economy, this is still often the case.

But you are lucky that these natural market factors are working in your favor, because in maturing tech markets, several other factors are conspiring against you. Unfortunately for most tech companies, what was once an innovation shortly becomes a commodity. Products gradually get cheaper and margins get smaller. Any computer company knows that. Any cell phone or TV maker knows that. But now even long-sheltered markets like software and networking are beginning to feel the squeeze. So how do companies respond? They put more focus on market share. They employ more discounting and promotion to obtain higher volumes. Then the competition responds with yet lower prices. Our company then follows suit. Margins drop even lower and the cycle repeats. Hopefully, your company has a proprietary technology or some other kind of market lock that makes you immune to this. But for most tech companies in maturing markets, this process is a growing reality. This is easy to see in consumer technology by simply picking up a newspaper. There you will see manufacturers fighting it out on price in four-color retailer ads. Other enterprise markets aren't on such public display, but their sales forces are doing more discounting and more bundling of products and services (with discounts on the whole package) as competition becomes more aggressive. Many tech markets actually get to the point where new customer acquisition is not only a low margin business but a *negative* margin one. As a result, the only people you can count on to make up the shortfall and eventually yield a profit are your existing customers.

Now the complexity avalanche is throwing an unwanted wrench into this well-worn process. Many customers—both new and existing—are being driven by the cost and confusion of the complexity avalanche to become pragmatists, not technology "romantics." Sure, they still love sizzle, but existing customers increasingly are asking why they need more features when they know they can get more value out of the model they already own. One large software company that was ready to release a major new software version in the summer of 2009 reportedly had to face the reality that only 8% of its customers had upgraded to the LAST new version since its release in 2007. Why do corporations need more servers when their current utilization is poorly managed and there are tons of opportunities for efficiency gains? Why do hospitals need the latest functional MRI or CT device when the three-year-old model they haven't yet paid off still has advanced imaging capabilities that the staff isn't using? Do consumers really need a new camera to solve their red-eye problem when the one they already own has a red-eye removal feature?

In enterprise and commercial technology markets, IT departments are becoming more resource-constrained every day. They have tighter budgets with smaller staffs to maintain an ever-growing, increasingly connected and complex stack of technology. In a recently released survey called "Top CIO Priorities for 2009," the Web site CIO Insight revealed that "discovering and deploying new technologies" was *last* among CIO priorities (number seven out of seven). The number one priority? "Improving or creating strategic applications"—in other words, getting more business value.[1] And what is the key to that? We would argue that the low-hanging fruit of improved business value from IT is around better end-user adoption.

In 2005, George Bailey and Hagen Wenzek wrote a book called *Irresistible*[2] in which they predicted that consumer electronics companies would have unending, exploding demand as consumers clamored for the next great gadget or device. But is that really true? There's no question that *occasionally* a product comes along—think iPod or iPhone—that does create "exploding demand" for the manufacturer. But for most products, it is a long runway and many never really get off the ground. As Geoffrey Moore has pointed out in his landmark series of books,[3] including *Crossing the Chasm*, there are certain people who find gadgets or features irresistible and will throw away a perfectly functional old model just to get the latest and greatest new one. He calls these people innovators and early adopters. But Mr. Moore goes on to make a key point: Innovators and early adopters are not a very big market. We agree. Yet you could make a very long list of "one-hit wonder" companies who IPOed by selling to early adopters and then never did much else. The product never established a reputation among the larger market of pragmatic customers. The problem is that none of those executives saw themselves as "that" company while they were in the middle of it. They didn't want their market to dry up after the early adopters. Some companies missed the chance to guide their marketing strategy properly. That is Mr. Moore's key insight. Others made the mistake of not wrapping the product in the service blanket that was needed to enable the larger market of traditional buyers to be successful with their leading-edge technology. Maybe they thought that either they could engineer their way out of the customer adoption problem or that their channel would eventually come to the customers' aid with services. When neither happened, the product stalled. They failed to deliver what Mr. Moore refers to as "the whole product."

Thus one of the most damaging repercussions of the consumption gap is that it actually reduces the size of any tech market where it exists. Too often products never become usable by the majority of potential customers. The classic example is how seniors were slow to embrace PCs in the 1980s and 1990s. Computers seemed so overwhelmingly complicated that older people stayed away in droves. Had Compaq and others addressed the gap, they would have had a far bigger market to sell to in the early years. Another example of the consumption gap shrinking a market is enterprise class software (CRM, ERP, etc.) for small and mid-sized businesses (SMBs). For years SMBs were slow to purchase the low-end offerings from SAP and Siebel because they thought it would be too complex or too pricey. This opening provided a ripe opportunity for Salesforce.com to come along and convince SMBs that it had a simpler solution to the problem. Getting products in better use, by more technically-challenged customers, and thus creating more value can expand any market to include segments that have been sidelined by the complexity avalanche.

This may not be necessary for your company if you have the skills to be able to constantly repeat having hit products that drive "techies" crazy. Not many companies do today. Most are mortal and have to fight hard to get the results they want.

Even if you were sure you could have a string of hits, how much more could you sell if you were successful with the rest of the market that are not early adopters? How many homes in the United States and Europe still don't have a wireless network? How many business-people still don't have a wireless broadband card that lets them access the Internet from anywhere that a cell phone works? Just do the math: How many potential customers are in your early adopter market when compared to the potential market of more traditional customers who generally find irresistible functionality quite resistible, maybe even intimidating?

So if you really want your company to be successful with main-stream (or, as Moore calls them, "majority") customers and all you have to offer is more bells and whistles, you could be leaving a lot of money on the table. We are not just talking about the laggards who will never buy; we are talking about the large group of pragmatic customers that need to be convinced that they will be successful with your new product and its new features. Among these customer pragmatists appetites are changing: They are not buying new technology just because it is there. As we mentioned before, many existing customers have slowed their upgrade frequencies because they're satisfied using the basic functions of their stable systems. *They have stopped growing as users;* they have stopped needing more. This is a key point that will effect your future revenues. Others now understand the true costs and hassles of the upgrade process, and they are demanding more vendor-supplied services to reduce those impacts. The effect on your available market cannot only be seen in the behavior of buyers. In enterprise technology you can also see how it is limiting your available market to just certain users inside your existing corporate customers. And for mature technology markets, this may be even a larger problem.

CHANGING DEFINITIONS OF *VALUE* AND *SUCCESS*

In Chapter 1, we introduced the consumption gap as being the difference between the value a product or solution *could* provide and what it actually *does* provide. But what, exactly, is "value" today? And while we're defining terms, what is "success"? Well, it depends on whom you ask.

Consciously or not, the tech industry has historically asked its customers to define success as the availability (not the use) of great product features. After closing the sale, we generally provide enterprise customers some professional services, which are designed to take customers through installation and integration. If it's a consumer product, we may provide customers a three-step, brightly colored, cardboard foldout in the very top of the box. The goal is the same: to make the product useable. With SaaS and cloud computing we can even allow customers to skip these early stage activities altogether. Once the product is up and running, our customer service and support organizations try to help the customer maintain availability for our product's valuable features.

We think that technology customers today, whether they're individual consumers, SMBs, or enterprise users, are developing a different take on success, consciously or not. In their minds, it's increasingly about whether the product's key features are *actually* being used and delivering their potential value. Have they truly achieved end-user adoption (EUA)? This is not only a "use or don't use" debate. This is about the right use, the best use, the optimal use. And EUA matters whether we are talking about an IT guy using a system management tool, a business user using ERP, or a consumer using a GPS. It seems a pretty reasonable request from a customer. But we as an industry have no way to answer this demand. We can promise to deliver product availability to the customer, but nothing more.

To better understand this problem in the enterprise market, TSIA recently partnered with Neochange and the Sand Hill Group to survey enterprise IT execs. Of these executives, 69% said the number one driver of value realization was "effective user adoption." Only 18% defined value as "functionality."[4]

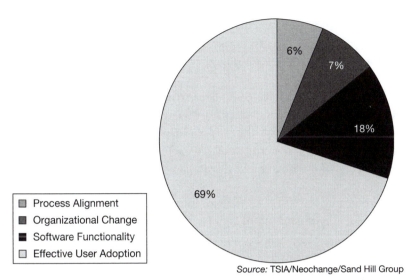

FIGURE 2.1 IT Buyers Overwhelmingly See Effective User Adoption as the Key to Achieving Value.

Taking this a step further, the vast majority of those executives said they view value realization—far more than product deployment or implementation—as the definition of success.[6]

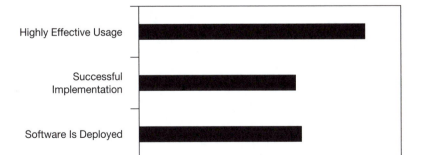

FIGURE 2.2 Effective Usage Is the Buyer's Main Measure of Software Success.

Dean Lane, author of *CIO Wisdom*, says, "The biggest challenge for the IT industry remains delivering value for the dollar. What makes this more important ... is that there is no next Killer App. ... So the challenge becomes how to derive value from systems that are already matured. This is not something that IT or functional users can do alone."[6]

This all sounds pretty basic, doesn't it? There's only one problem: For the most part, end-user adoption isn't nearly as successful as it should be. Research suggests that this is one of the main reasons that as much as two-thirds of IT investments fail to achieve their intended business results. Let's take a look at a few examples.

- According to our survey, 50% of software functionality paid for and licensed by organizations is not actually used.[7]

- A whopping 83% of SMB owners in a recent AMI Partners survey say that getting staff to use the software was their biggest CRM challenge.[9]

- Of health care organizations that have computerized physician order entry (CPOE) capabilities, only 19% have physicians entering 50% or more of their orders. And almost 60% have physicians entering less than 10% of their orders.[10]

End users not getting value out of their products has some pretty serious consequences. On the most basic level, products that don't get used don't deliver value. In many cases, that missing value can actually be quantified. In health care, for example, a number of studies indicate that CPOE can have a significant economic impact. Each major CPOE user could potentially save a health care organization $100,000 or more per year. A typical 500-bed facility could realize savings of more than $10 million per year—but only if physicians use the technology.[11] There's also a definite link between adoption of CPOE systems and patient safety. The Institute of Medicine attributes between 44,000 and 98,000 deaths per year to medication errors, and believes that CPOE can be a powerful tool for preventing such errors.[12]

One big problem that enterprise tech companies are now awakening to is that the industry has been selling its products to buyers who themselves underestimate the importance and cost of end-user adoption. You can most easily see this in customers' IT budgets. Whether buying new products or extracting more value from existing ones, IT budgets consistently underfund end-user adoption costs. You could argue the chicken-and-egg scenario here: What comes first—tech companies offering EUA services or tech buyers allocating money toward it? What is clear is that more and more IT executives are recognizing that something is not right.

Didier Lambert, the CIO at multibillion-dollar optical lens manufacturer Essilor International, put this in an even simpler context. At TSIA's recent Technology Services Europe 2009 Conference, he said: "If I could just get one message across to all of you (technology companies) it would be this: As a CIO I am not the customer. I am just a reseller for you. The real customers here are the business users. Your success and mine depends on how happy and successful THEY are, not me. But you keep acting like I am the customer."

In fact, 88% of respondents in our survey of IT execs say they now expect value enablement and/or process alignment services from their vendors. Only 12% said they were satisfied with the provider offering them product expertise only.[14]

What Software Buyers Expect from Their Providers

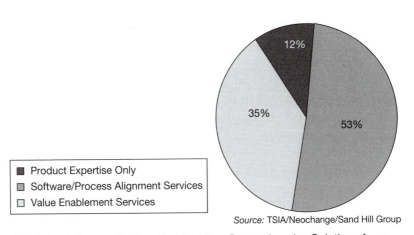

Source: TSIA/Neochange/Sand Hill Group

FIGURE 2.3 Buyers Are Looking for More Comprehensive Solutions from Their Providers.

In addition to *demanding* services from their vendors that enable true value, corporate customers are now starting to actually *measure* it—and that should strike fear into the hearts of more than a few tech companies.

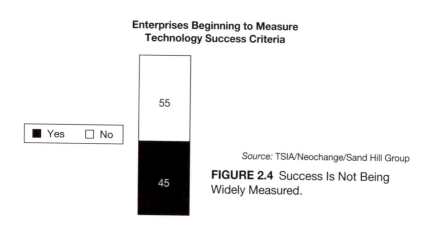

Enterprises Beginning to Measure Technology Success Criteria

Source: TSIA/Neochange/Sand Hill Group

FIGURE 2.4 Success Is Not Being Widely Measured.

While many are doing this using manual models and surveys, a growing number of customers are beginning to gather hard data about end-user adoption. How? With even more technology.

Supporting our own TSIA findings, Forrester estimates that by 2010, end-user monitoring will have penetrated 40% of the enterprise industry.[15] A study published jointly by ECPweb, Macrovision, Soft-Summit, CELUG, and EDA Consortium went even further, finding that in 2008, 58% of enterprises were using automated software management tools (up from 44% in 2006). Although much of this tracking is being done to ensure compliance with vendor agreements, 44% say their primary reason for doing it is to reduce software costs (up from 31% in 2007).[16] Neochange suggests that "the improved usage transparency this technology provides is also clearing out 'shelfware.' In fact, according to ServiceSource, one of the industry's largest maintenance contract sales outsource firms, 30% of software maintenance contract cancellations are attributable to lack of effective adoption."[17]

One particular type of software-tracking software is application mapping (AM), and you'll be seeing it at an enterprise near you very soon. Gartner predicts that AM and automated software license track-ing will have 90% penetration by the end of 2009.[18] Both of these new technologies will give enterprise customers detailed insights into their primary success metric: end-user adoption. Forrester Research estimates that AM software will reduce enterprises' application costs by 10% to 30%—most of which will come at the expense of vendors.[19]

What is the bottom line for tech companies? The handwriting is on the wall: Something radical must be done to overcome the com-plexity avalanche. It has and will increasingly become a gating factor in the growth of the entire tech industry. We believe that "features and availability" will devalue (but certainly not go away!) as the measuring stick of IT vendors, and end-user adoption will emerge as a new and defining capability.

CHANGES IN HOW THE MONEY GETS MADE

If you were to ask the sales and marketing departments—or the CXOs—of most tech companies whether they're a products or services company, the vast majority would insist that they're in products. It's where they came from, it's what they know, and for decades it has been the proven path to wealth for entrepreneurs and shareholders alike. Our argument is a challenge to this conventional wisdom: For the first time many of these companies—whether they produce enterprise software, hardware, industrial equipment, medical devices, or consumer electronics—may be wrong both financially and strategically.

At TSIA we track something called the "Service 50," which is an index of today's 50 largest technology solution providers. Some are "independent" services companies like Accenture or Wipro, but most are tech product companies with imbedded service businesses. Together these 50 companies account for over $800 billion in total revenue (prod-ucts and services combined).

Gartner recently estimated that in 2008, tech and telecom services globally accounted for a stunning 70% of the $3.4 trillion in global IT spending. IT services alone were 24% ($872 billion).[20] And the fastest growing part of the technology market (in total dollars) is not products but the *services* customers need to operate them. The services sector is growing over 10 times as fast as hardware markets. (Software has a similar growth rate, but at $211 billion in annual sales gross dollar spending growth on IT services leads software 4:1.)[21] That makes the tech services market far larger than the entire consumer electronics industry, and larger than the entire software marketplace, and it is likely to grow at a much faster rate than just about any other broad product category. You get the point: Services is already a pretty big market and it is going to get bigger. The question for you is whether to jump into what may be unfamiliar territory or remain committed to being a product-only company and hope a successful service ecosystem develops around your technology—one capable of driving EUA.

Looking at typical large enterprise software companies, the evidence is pretty clear about which option is more lucrative. For these companies, an average of about 60% of total revenues now come from service, support, and maintenance (that's up from under 35% in 1999). Plenty of companies are actually higher than that. Oracle and SAP, for example, bring in close to 70% of their revenues from services and maintenance.[22]

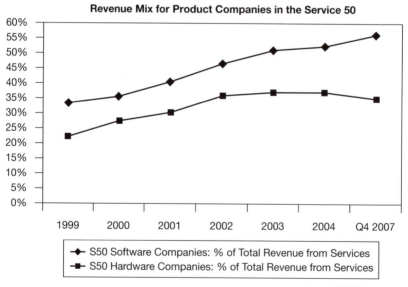

FIGURE 2.5 Shift in Product-Service Mix.

Even if you carve out a healthy portion of the maintenance revenue for software companies and re-classify it as product revenue to compensate for the annual product update, it is still a huge number. In hardware and systems, the average revenue from services is nearly 40% (up from just over 20% in 1999). But even here there are many companies for whom services revenue are far more important. For Unisys, its around 90%, IBM is 60%, and Xerox is 50%.[23]

Of course, revenues don't tell the whole story, so let's look at combined software and hardware margins for product companies in the Service 50. Between 2005 and 2007, the average margin on products declined from 68% to 62%. Over the same period, the average margin on services increased from 48% to 53%.[24]

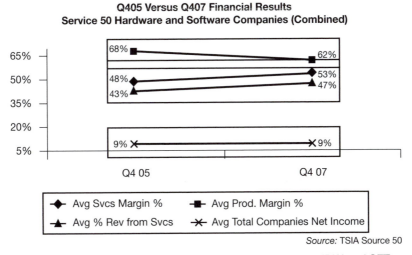

FIGURE 2.6 The Global Service 50 Shift in Margin Dollars: HDW and SFT.

That margin gain from services was enough to offset eroding product margins and allow overall net income to remain flat. Without those service gains, many a stock price would have tanked.

The reality is that the main future source of gross profits is not going to be from the product. Product margins are on a death march in many tech sectors. For hardware companies, the window to make money off a differentiated product or feature is shrinking. Competitors can turn your innovation into a commodity at astounding speed. That makes the period for high margins due to that differentiation very brief. How long can Intel charge a premium price for a new chip before AMD is offering an alternative at a lower price? Two or three months? That is true in the systems businesses as well. And with SaaS and cloud computing models gaining market share, even the software business is headed in this direction. If Oracle puts a great new feature into its CRM module, Salesforce can replicate it and have it in use by customers extraordinarily quickly. And slowly but surely, low-priced software solutions are springing up in market after market.

So, being the product companies that we are, how do we react? Many do what they know: Add more features faster. And then guess what? With each new feature cycle the consumption gap grows. The complexity of the product increases and the percentage of features in actual use decreases. This is not an effective strategy to address the problem.

More and more, the money will be made in services. Do you want to see how real this story already is today? Following is a financial analysis of two tech companies derived from publicly available financial information—a software company (Oracle), and a storage company (NetApp). The Revenue and Cost of Revenue allocations come from public financial statements. We document our allocation assumptions in the model itself.

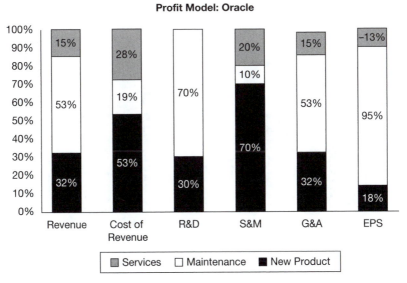

Profit Model: Oracle

Legend: Services | Maintenance | New Product

ASSUMPTIONS:

1. Revenue and COR taken from Oracle FY09 10K.
2. R&D expenses are allocated 70% to maintenance revenue and 30% to new product revenue.
3. Sales and marketing expenses are allocated 70% to new product revenue, 20% to services revenue and 10% to maintenance revenue.
4. G&A expenses are allocated based on percentage of revenue.

Source: 10K FY09

FIGURE 2.7 Maintenance Is the Main Profit Engine for Large Enterprise Tech Companies.

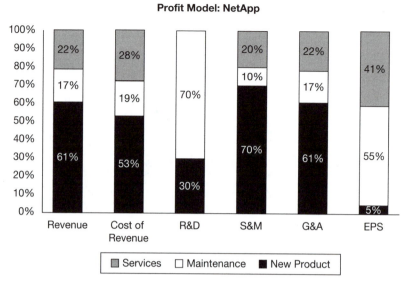

Profit Model: NetApp

ASSUMPTIONS:
1. Revenue and COR taken from NetApp FY09 10K.
2. R&D expenses are allocated 70% to maintenance revenue and 30% to new product revenue.
3. Sales and marketing expenses are allocated 70% to new product revenue, 20% to services revenue and 10% to maintenance revenue.
4. G&A expenses are allocated based on percentage of revenue.

Source: 10K FY09

What is clear is that services are not only the fastest growing segment of many tech markets, but they are also a huge driver of profitability.

So … if you are running a company with 30% to 60% of your revenue coming from services (and growing) and 30% to 90% of your net operating income coming from services (and increasing), can you really continue to believe in your heart that you are not in the service business? Not that size alone makes a line of business strategic, but the complexity avalanche is making your company's ability to provide new services not only financially important but critical to the success of both customers and your product business. And this same complexity is changing the order of the business model. The services will be pulling through the products, not vice-versa.

WHO WANTS TO BE THE TRUSTED ADVISOR?

For this reason alone, you may want to become even more committed to services because those services will end up heavily influencing your product's market share.

Let's face it, the consumption gap has spawned a huge and rapidly growing industry of third-party organizations that promise to do what manufacturers can't or won't do: Help customers get the most value out of their purchases. They are not only stealing revenue from manufacturers but stealing something else that's equally important: the customer relationship.

To understand this issue we need look no further than the relatively simple world of consumer electronics. Big-box retailers have been the latest to step up to the challenge of offering third-party services. Best Buy's Geek Squad and other consumer service offerings have hit the market with a bang over the last couple of years. These product retailers had more than one motivation to enter the service market. The first was to close the consumption gap. There is no doubt that the lack of available services and training was causing customers to return products—not because they're defective, but because they couldn't get them to work. Those returns are exceedingly expensive. The second is to stimulate new revenue sources with superior margins to selling the products themselves. In fact, the business of selling technology hardware to consumers gets worse all the time. Between competitive manufacturers and competitive retailers, making a profit just selling boxes is nearly impossible. Just ask Circuit City. Services offer more attractive margins and thus are catching the interest of big-box retailing executives who see a real chance to supplement low product margins by wrapping them in higher margin services.

But besides just reducing product returns and getting a more profitable revenue stream started, these companies are starting to realize the true power of providing services that add value to the customer. Say you're at Best Buy and you've just picked out a brand-new home theater system. It looks pretty complicated, so you decide to pay an extra $200 to have the Geek Squad come to your home and set everything up. A few days later, Bob from the Geek Squad arrives, asks you about how you plan to use your new system, and gets it up and running. He trains you, your spouse, and your kids on how to use the remote control. Since Bob is there, you ask him a few questions about that home wireless network you haven't been able to get to work, or that new printer that refuses to cooperate with your equally new laptop. Bob, being a nice, knowledgeable guy, is more than happy to help you out with a few good suggestions. By the time he leaves, he knows a lot about your situation—how many computers you have, where they are, what you use them for, and what your biggest concerns are.

A month or two down the road, you decide to add a wireless printer to your home network. The number of choices and options is overwhelming, so you think about asking someone for a recommendation. Whom would you like to pick up the phone and call? Bob, right? After all, he knows you. And because you trust Bob, chances are whatever he suggests is what you'll get. There's no sense spending hours comparison shopping on the Internet or scouring the Sunday paper for sales. You'll just drive down to Best Buy and pick up the recommended equipment. Bob's suggestion might not be the cheapest in the category or the one that's on sale, but you'll buy it anyway. And that's good news for Best Buy's product margins.

According to a 2007 report by the Consumer Electronics Association and ServicesRevenue Inc. entitled *Beyond Delivery and Installation: Premium Services Consumers Want,* consumers are doing just that: "When retailers offered services such as installation assistance, the initial motivation was to reduce return rates or generate incremental revenue at a decent margin. The problem for retailers is that when the installers did a great job ... the consumer kept calling them."[25]

In the enterprise space, the same basic thing happens, except that Bob is from Accenture or Bearing Point. Say you're the CIO of American Express. On any given day you'll have sales reps from Symantec and McAfee and other software companies knocking at your door with security solutions. A few years ago you might have gone with the vendor who offered the most features. But today, with dozens of systems in place, running different applications from different vendors, you've brought in Accenture to ensure that everything fits together seamlessly and transparently. And when Accenture recommends that you standardize and integrate all of your security software companywide, you'll likely look favorably on whatever vendors and systems it recommends.

While the rise of the trusted, third-party advisor is fine for customers, it's potentially dangerous for manufacturers. If you're Sony, for example, and you're selling your Blu-rays through retailers like Best Buy, your product business might become dependent on what the Bobs of the world are saying to their customers. Sure, your huge spends on advertising and brand awareness and strong word of mouth are worth something, but when it comes down to pulling out the credit card, your marketing department may be no match for Bob.

If you're Symantec or McAfee, it might be Accenture, not Amex, who's going to determine who gets the deal. And if you're VMware competing with Microsoft to land a $20-million virtualization deal at Bank of America, you might be dealing with Capgemini or whoever B of A's trusted advisor is.

So whether it's the Geek Squad on the consumer side, or Accenture and Capgemini on the enterprise side, third-party providers are proving they can build big businesses out of servicing the complexity avalanche. Ultimately, the greatest asset these companies are building today may not be their finances or their capability, but their end-customer relationships. As these companies aggregate trusted advisor relationships with your customers, they not only disintermediate you, they could even begin to act as a "group purchaser" with a lot of price-negotiating leverage. You might have fewer end-customer relationships and your salespeople might spend all their time across from huge, influential third parties with massive buying power. About two years ago TSIA partnered with some Wharton Business School professors to study likely long-range scenarios for the technology industry and its service ecosystem. This incredibly insightful TSIA report, called *Navigating Uncertainty: Future Scenarios for Technology Services* correctly identifies this type of market aggregation in the enterprise market as one of four highly likely scenarios to emerge.[26] For most consumer product companies that day is already here. They know the massive buying power of the retailers and the impact it has on their product margins. What happens if Staples and Office Depot begin aggregating SMB relationships and Accenture and Capgemini continue to aggregate large enterprise relationships? Where will your high-margin segments be if the most profitable customer relationships are all owned by a handful of large service providers? What if cloud computing takes off and there become but a few, very large server customers who are building and maintaining the cloud that thousands of your former server customers are now using?

Another compromise that companies who service through channels face is a real loss of customer intimacy, insight, and data. How are customers *really* using the product? What do they *really* want the product to do? What new features do they want? What new opportunities are emerging for next-generation products and services? And there is an even more fundamental question: Can you really control your own customer satisfaction if the increasingly important service relationship moves out into the channel?

Not surprisingly, while most manufacturers are slow to truly integrate services into their "product, product, product" mantra, there are a number of interesting experiments going on right now. As we mentioned, we are paying particular attention to Apple's Genius Bar, where for $99 Mac buyers can get a year of weekly one-hour private lessons with a Mac expert (a Genius) focusing on whatever the customer feels would help him or her get the most out of his or her computer. Not only is this service selling extremely well, it's also creating brand loyalty and referrals that are the envy of the entire industry. Services have become a real part of Apple's strategy. Other consumer tech companies could opt for a similar model. In fact, Microsoft has recently joined the fray by announcing plans to put trained "Microsoft Gurus" into retailers.[27] The gurus will answer customers' questions about Microsoft products and PCs in general, and will give demos on how customers can get those products to work together. Obviously, the hope is to get the customer to keep the Microsoft platform front of mind. As we said, third parties like the Geek Squad may have initially sprung up to fill the void created by manufacturers' underinvestment in services. But now Best Buy sees services as a cornerstone of its future growth and profits,[28] including expanding the Geek Squad into the SMB market. The interesting question now is how manufacturers who sell through the channel rather than through their own stores will address this rapidly growing consumer demand. Will they capture any of this lucrative market or will it all go to third parties?

What all these early-warning signs should be telling you is this: In a sense, you don't really have a choice. If you're a software company, a network company, a device company, a medtech company, a Web site service, or have nearly any other role in selling complex technology products to end customers, the question isn't *whether* your customers will demand more services; it's *when*. The complexity avalanche is going to drive your customers into some service provider's arms. Who are the Bobs of your business? Are you really prepared to have them become your customers' trusted advisor? If so, how are you cultivating that community to behave the way you want it to? Or do you want your company to assume that role? If so, what is your strategy to win that relationship by having the best service solutions? While these are age-old questions about selling through channels they are vitally important questions to consider as we move on in our discussion.

SERVICES' IMPACT CAN BE MEASURED

It is easy for us to see the revenue and profits that come directly from a service offering. But the more important impact of services may be from something we don't even measure today: Product pull-through. Calculating the total economic impact of services has been the holy grail of the service industry for a thousand years. It is an evolving model and one that TSIA is working to perfect. But we do have some case data, like the following analysis of account profitability at a public software company.[29]

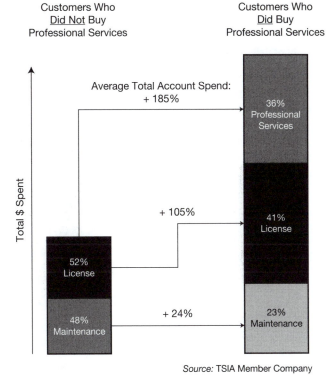

Effect of Selling Professional Services on Total Account Spend

Customers Who
<u>Did Not</u> Buy
Professional Services

Customers Who
<u>Did</u> Buy
Professional Services

Average Total Account Spend:
+ 185%

36%
Professional
Services

+ 105%

41%
License

52%
License

Total $ Spent

+ 24%

48%
Maintenance

23%
Maintenance

Source: TSIA Member Company

FIGURE 2.8 Comparison of Total Spend on Software and Services.

The company wanted to understand whether including professional services in deals actually had an impact on how much money the
customer spent on products later on. To do this analysis, it did a very
simple but revealing exercise. It separated those customers who took the
"customer as general contractor" approach and handled the installation
and implementation of the products on their own or with a third party.
They did not contract for any professional services from the company.
The company then computed the additional purchases of this group of
customers over time and used it as a baseline.

In this case, this group of customers did not buy any advanced
professional services. The size of the bar represents their relative spend
on each category.

By contrast, those customers who did purchase advanced services
had a total revenue spend profile that was 2.85 times as large as the
service light group. This included a 74% maintenance renewal rate
(nearly 50% higher) (see Figure 2.9 below).

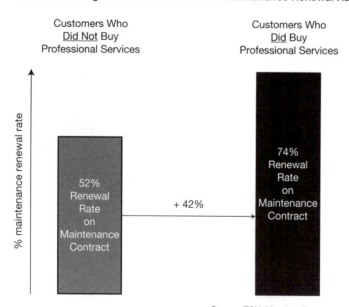

FIGURE 2.9 Comparison of Maintenance Renewal Rates.

Another public software company wanted to understand whether or not there was a relationship between the provision of services among its highest growth accounts and its lowest growth accounts. By comparing these two customer groups it discovered an amazing relationship.

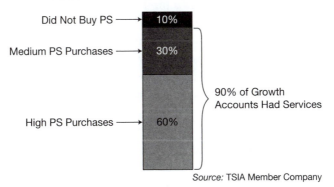

FIGURE 2.10 Use of PS by Customers with Growing Product Revenue

Among the company's top revenue growth accounts a whopping 90% were either medium or heavy spenders on professional services. Only 10% were going it alone or exclusively with third parties.

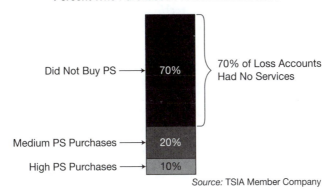

FIGURE 2.11 Use of PS by Customers with Declining Product Revenue

By contrast, those accounts whose amount of purchases dropped the most in the same period had a startlingly different appetite for services. A stunning 70% of these shrinking violets had not purchased any professional services from the company and only 10% were heavy spenders on services. This data was very influential on both the executives and the sales force. They learned that there was a direct correlation between attaching professional services to a product sale and the loyalty and revenue that resulted from that customer.

In a similar study shared at our Technology Services World conference, a major hardware systems company found that approximately 94% of their PS accounts either maintained or improved their relative core product bookings volume. That was nearly 50% higher than among the non-PS accounts.[30]

These companies are among many who are rapidly building their internal business case for change. All the companies that we know of who have taken the time to do some analysis have discovered the same phenomenon: Services sell more products.

THE "LAWS OF SERVICE GRAVITY"

The purpose of this book is to explore and analyze existing service business models as well as propose future ones for tech companies across a variety of industries. Fortunately, since tech services have actually been around for over 50 years, we can look back on several distinct phases in its history. At TSIA we believe there are some intractable laws about the servicing and supporting of tech customers that have always been present. These laws of service gravity can help us speed through the inevitable debate that this book will create inside your company and even across the industry.

We believe there are at least five inescapable laws of services gravity for tech companies.

1. Complexity forces every technology company to move into the service business.

2. Technology services become more profitable over time. Products do the opposite.

3. Product features give rise to a need for specific new services but they have a shelf life. These services are ultimately "sunsetted" back into a future generation of the product ... and a new service requirement cycle begins.

4. Because of Laws 1 and 2, the value and the margins in individual tech markets inevitably shift from the products to the services that enable it.

5. The "end state" of tech offerings will almost always be as a service.

The First Law of Services Gravity:
Complexity forces every technology company to move into the service business.

As James Womack and Daniel Jones point out in their great book *Lean Consumption*, a customer has to search for, obtain, install, integrate, learn to use, maintain, repair, and eventually dispose of every product.[31] For many of these customers, the cost of engaging in these activities far surpasses the cost of the product itself. It is also true for the tech companies who make the products. They spend more to provide assistance (services) to the customer throughout these various stages than they do to develop and produce the product. And the more complex the technology is for end users, the greater the imbalance. Thus we maintain that tech companies that sell complex products to end users *must* evolve into service companies over time. Their customers demand it and their investments should reflect it. This is not a "decision" that tech companies can choose to make; it is the first law of service gravity. The concern is over how much time and how many missed customer opportunities it will take before senior management accepts this reality and begins to think about services from a strategic perspective rather than just as an add-on to their product business.

The Second Law of Services Gravity:
Technology services become more profitable over time. Products do the opposite.

Services evolve along a well-worn path. This path includes three distinct phases:

1. The Age of People
2. The Age of Tools
3. The Age of Product

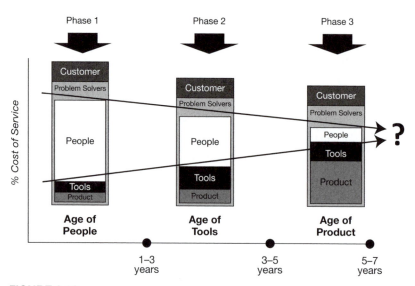

The Phases of Service Development

FIGURE 2.12

In the Age of People, your employees are out struggling to gain new knowledge with your customers. If we use today's modern customer support delivery model as an example, we would say that this age was in the 1970s and 1980s. A lot of software support was not automated and much of it was even delivered onsite. Knowledge came and went in and out of support people's heads with little institutional benefit. By the 1990s we began to systematically apply tools to the problems. The net effect of this Age of Tools was a reduction in the number of people needed to deliver customer service. Finally we entered the Age of Product for support where the products themselves are being built to self-heal, avoid problems, and seek their own solutions remotely. This will further decrease the number of people needed to support the same number of customers. So in this way tech services tend to become more efficient over time. If it is a revenue-based service it also means that the gross margins from services will improve.

The Third Law of Services Gravity:
Product features give rise to a need for specific new services but they have a shelf life. These services are ultimately "sunsetted" back into a future generation of the product ... and a new service requirement cycle begins.

The reality is that, like products, services have their sunrises and sunsets. The services we must develop now to address the consumption gap will be our next service sunrises. They will move through their phases and eventually we will sunset the services into the product. Eventually most tech services become so imbedded into the product that they cease to be visible to the customer. This can be seen in today's maintenance service business. With self-healing systems, user community support networks, preventive maintenance, and remote support tools, an enterprise customer might not ever see the support and maintenance that their vendors provide. While objectively that is a good thing, from a value proposition standpoint it can actually be a problem. At TSIA we call it "the invisibility of great support." If support margins begin to decline, it will be a revenue-side collapse, not an efficiency-side one.

But don't worry, there will be new problems or opportunities and new services needed to address them. That is the nature of the complexity avalanche. And such becomes the ironic dance tech companies must perform: The products give rise to a need for new services and then the services are ultimately sunsetted back into the product. And then a new cycle begins again. It is a phenomenon that is here to stay.

Our next generation of technology adoption services will also go through this cycle, from The Age of People to the Age of Product. Fortunately the cycle times will be shorter because they will be informed by our recent experience with state-of-the-art tools and processes. But it will still initially be labor intensive. New knowledge will need to be created, and new tools defined, built, and implemented. Eventually your new services organization will become the product development team's best ally—if it isn't already. It will be the number one source of product enhancements, supportability improvements, and ease-of-use suggestions. Many of the tools created by the services organization may go directly into the product.

The Fourth Law of Services Gravity:
Because of Laws 1 and 2, the value and the margins in individual tech markets inevitably shift from the products to the services that enable it.

The revenue and the profits in tech markets start in products. But as we already mentioned, they quickly commoditize. And this occurs faster and faster every year. If it weren't for the complexity avalanche a tech product category would quickly be all done. There would be no profitable activities on the horizon for the tech company, just more efficiency drives and more cost cutting as a way to shave a tenth of a point of product margin here and there. Are you really going to win that game against a competitor from China or India? No. But because every digital technology ultimately causes a consumption gap, people become dependent on service providers to help them to achieve the product's full potential. And so the tech company wisely begins to shift its focus toward the services and tools that can close the gap. Remember what Gartner said, that 70% of all enterprise IT spend is on the telecom and the service to make the technology work, not on the products themselves. So as the company with the most knowledge of our own products and how they get used, why wouldn't we capture our fair share of that pie? Not only is there more revenue there but the margins are higher than the product and they get better with volume. That's what companies from IBM to Best Buy have figured out.

In a 2007 analysis of the enterprise software sector, Chasm Institute and Neochange wrote, "... *[T]he need to grow revenue through services becomes not only an option; in fact, it's an imperative. But what previously worked in a product-centric market doesn't guarantee success in a services-centric world. Moving forward, services innovation is likely to provide ISVs (Independent Software Vendors) with the highest returns. By incorporating an effective user-adoption objective, software providers can focus their transformation and develop a new foundation for market leadership.*"[32]

Even some of the biggest holdouts in the "product is everything" world are beginning to realize that the service era of old is changing. Cisco, for example, which is famous for handing most services to its channel partners, has been talking about services-led selling for its complex system products like Telepresence. For these important and complex products, the company is actually bringing more service functions back in-house.

The Fifth Law of Services Gravity:
The "end state" of tech offerings will almost always be as a service.

The fact is, tech should probably have emerged as a service offering from the beginning. Leading companies today are realizing that the idea of selling a tech product as a transaction is shortsighted. Selling customers a tech product is and should be the beginning of a relationship that—if executed well—might never end. The tech company's part of the relationship is to be out there doing what it does best, driving new capability into the product. Then (and here is the part we don't do) it should help the customers drive that capability into true business (or personal) value. At the end of the cycle, it is time to start again. The customers are ready for more capability and that is exactly what the company has ready for them. They expand their product purchases and the tech company makes sure those follow-on purchases result in additional business/personal value. And it goes on and on until the product cannot be improved or replaced by a better one. Customers should not care about the product, they should just care about (and pay for) the results.

When I think of the how the tech business could have evolved, I always think of ADP. It was the original SaaS/outsourcing/managed services company. Then, for a while, payroll software applications for small business took market share from it. But in the end, the complexity avalanche reigned supreme. Payroll (with its tax, legal, benefits, reporting, exception, and other continually changing requirements) turns out to not be so simple. ADP is not only back as a technology solutions provider, it is dominating its original market of 50 years ago. And you pay by the employee per month, not by buying a $300,000 app.

An offering based on results is a dynamic, always-evolving service relationship. It is a service relationship that will pull through more products and/or revenue. That is how tech probably should have evolved. Clearly it is the direction we are moving in with cloud computing and SaaS.

We believe the rise of services in tech is upon us. And it's not for the same old services in the same old way. You may hold the deep belief that you must keep focused on your core (products) and outsource what is context (services). But think of it this way: By becoming a services company you'll become a more successful product company. IBM, HP, Apple, Oracle, GE, Xerox, Philips, and many other leading companies—large and small—have already begun the journey.

In Chapter 4 we'll talk about how you can dig your customers and your shareholders out of the dangers of the complexity avalanche. You'll learn how you can begin to close the consumption gap for your customers using the business infrastructure you already own. But before we do that we need to take an important look at the major service business models of today.

3 | Growing Problems with Today's Tech Services Business Model

Before we begin to talk about a new business model for technology service, we need to make an important point about the existing ones: At most tech companies today, the services that are being offered too often aren't the right kinds of services. And the ones that *are* the right kind aren't being offered in the right way. In both cases, the market is giving us unmistakable signals that it wants change. Change is needed not only to help drive the consumption of customer value as we have discussed but, even more urgently, to help defend your current post-sale business model.

Technology services can be grouped into four main categories. (Note: While there are other industry offerings like SaaS that could be called technology services, we are limiting our focus to service delivered around tech products, not where access to the product itself is the service.)

- **Customer Support and Maintenance**, which includes on-site field service, remote technical support (phone and Web-based), depot or retail returns, as well as software upgrades.

- **Professional or Consulting Services**, which includes all project-based offerings created to help customers through design, installation, integration, and implementation. Also in this category are services that help customers re-engineer internal business processes around technology.

- **Customer Education and Training**, which can be delivered in a classroom, a retail store, through the product, or online.

- **Paid-to-Operate Services**, where customers pay the tech company to take responsibility for some or all of their technology results, including outsourcing and managed services.

Let's examine the curent state-of-the-market in the two largest service categories; maintenance support services and professional services. For over 40 years, the profitability of the service line called "customer support/service and maintenance" has been a beautiful thing for tech businesses—ever since copier companies and mainframe hardware companies began putting their customers on annual maintenance programs. Over the years—and in particular since 2000—companies have employed technology, global labor arbitrage, and other innovations to make this a very high gross margin activity. For that reason, many companies that serve enterprise or commercial markets already recognize that the customer service and maintenance business is their most economically important line of business. But it also tends to be the most under-strategized line of business in the company by tech CXOs. At TSIA we refer to the customer service and maintenance business as "the financial engine and the strategic caboose" of most enterprise tech companies. But watch out for that thinking! Today some key trends have emerged that are threatening long-held, long-cherished assumptions about the profitability of the B2B customer service business. After we examine that, we will take a look at the quirky world of consumer services.

TROUBLE IN PARADISE

Corporate customers are now questioning the fundamental value proposition of customer service and maintenance contracts. This is but one indication that, as a line of business, maintenance services are probably in the "Decline Stage" of their Product Life Cycle.

For most enterprise IT organizations, vendor maintenance contracts are the second biggest IT expense after internal labor costs. Companies have been pounding away at reducing internal IT labor expenses for years. And a few years ago, encouraged by market analysts from Gartner and elsewhere, they began staging a rebellion against high-priced maintenance and service contracts. Many customers can't actually stop buying those contracts, but they resent it more every day.

There are two main reasons why they are locked into renewing their contracts. Mostly they need some guarantee of assistance if their products "go down." If an ATM network goes down, that is a problem. If an e-commerce site can no longer process orders, that is a problem. If a heart-monitoring device isn't working right, that is a problem. Even if your BlackBerry flakes out it can be a serious drag. So customers buy what amounts to an "insurance against downtime policy" from the tech company. If their system goes down, they know the tech company will get them back up ASAP. But even this insurance is amounting to little more than table stakes today, hardly something you can differentiate on. The second reason they feel compelled to renew their maintenance agreement is that they may want access to new features in future releases of software. Many enterprise tech companies smartly tied that to being "on maintenance." Their goal was (and still is) to make maintenance an "automatic" purchase by customers. But all is not perfect.

Following the go-go IT days of the 1990s, the big run up to Y2K, and the economic slowdown that followed, many businesses lost some of their desire for IT change. They started to see virtue and low expense in stable systems and processes—and for good reason. Today, many systems have few problems with downtime. As a result, some companies are delaying upgrades, doing at least part of their maintenance in-house or using low-cost third parties. After all, if you have stopped increasing demand on your stable systems, you're upgrading less, and since you need to drive more cost out of your IT budget, reducing your spend on vendor maintenance contracts seems like a good place to start.

Price pressure on hardware service contracts is nothing new—it's been around for over 20 years, ever since something called *multivendor services* became a reality. (That means that HP isn't the only company that can provide parts and labor to fix HP products. IBM can, and so can Unisys and about a hundred other third-party service providers, large and small.) This intense services competition is one problem. But couple that with the fact that every successive generation of hardware technology does a lot more for a lot less money, which has dragged product prices down. Since hardware maintenance contracts are often tied to list product prices, they get pulled down too. So declining hardware maintenance prices are a problem the industry has battled for a long time. What's different today is the amount of pressure customers can exert to force discounts and the level of sophistication in their tactics.

The story on the software maintenance agreement is slightly different and, in some ways, far more troubling. We believe the root of the problem is that enterprises have historically had a hard time quantifying the value of their software purchases. Today a number of lower-priced, lower-function products are beginning to win out over their older, more expensive competitors. They owe at least part of their success to their competition's inability to quantify the value that their higher-priced products offer. In other words, if you can't specifically quantify your value, you leave yourself open to lower-priced, (sometimes) lower-featured competitors.

The haziness around realized value is pretty much the same when it comes to lucrative software maintenance agreements. Enterprises have generally accepted the notion that insurance against downtime is worth money. And clearly it is. But what if they could get the same protection from software companies that operate on a 25% maintenance margin as they get from those that operate on a 75% margin? One company we know of has reduced their cost for "remote solves" from $1.20 per minute to $.30 per minute by moving these operations to the Philippines. Don't you think the customer's know that? Don't you think they want their costs to go down too? Of course they do. Software maintenance has never had to face actual competition before. But you can bet that the more mature the industry gets, the more alternatives to OEM direct service and support will become available. Unless software maintenance begins to be linked to value *creation* instead of value preservation, downward price pressure will continue to mount.

Clearly, the tech service industry's ability to justify the old uptime- and availability-based maintenance value proposition and price is getting harder today. This is true despite software updates being bundled inside the offering. And because the maintenance/customer service business is so huge and so profitable, this breakdown should be scaring a lot of people (if it isn't already).

According to TSIA data, about 16% of tech maintenance contracts today aren't renewed.[1] Actually, that is still pretty acceptable and seems to be true in both good economic times and bad for some product categories. So the 16% is not the bad news. The main way corporate customers have of fighting back today is not by cancelling contracts but by demanding discounts.

TSIA data also proves that corporate technology customers are, on average, able to successfully negotiate price reductions in one of every three service and maintenance contracts on their technology. A lot of these customers threaten to cancel their contracts or give them to another vendor. While few actually do, the magnitude of their potential cancellations is so huge they are emboldened to exhibit brinkmanship that can weaken the knees of most vendors. As a result, vendors often agree to deliver the same levels of service and/or longer contracts for less money.

Making everyone's life tougher today are the sophisticated new processes that enterprise and commercial customers use to purchase and renew customer service and maintenance contracts. Remember the good old days when we sold service to the customers we actually serviced? Back then, the people sitting across the table had become friends with some of your employees. They knew your support service was great because it had saved their bacon time after time.

Those days are gone. Now the people sitting across the table from your service sales person aren't IT people and they don't know or care much about the work you do. They're the purchasing department, and their job is simply to get the price down. Some come to the table with lawyers. Some come with special consultants who have developed a practice around renegotiating service contracts. They have reports from analysts telling them the best ways to pressure you. They might use a sophisticated third party to negotiate their RFP. They have data to prove that you are overpriced compared to their other vendors. They know how to tie the maintenance pricing to additional product purchases and to mention that on the last day of your fiscal quarter. And they have gotten pretty good at convincing vendors that they might just walk away from the whole relationship. Overall, they're damned good negotiators—often far better than your service sales rep. At the very least, they are going to try and run their demands up the chain of command, slow your renewal cycle, and drive up your cost of sales tremendously. Some of your biggest customers might go all the way to your CEO with their demands for maintenance discounts.

And guess what? It's working. Vendor discounting of maintenance and service/support agreements are on the increase. With the exception of companies that hold virtual monopolies in their industry segments, discounting has become almost rampant. After all, some software companies are making 90% gross margins on support and maintenance. We have been the beneficiaries of customers' inability to quantify value. Sophisticated customers have identified this disjuncture either intuitively or with hard data.

Just to give you an idea of how this plays out on a larger scale, consider what happened to one major office solutions company. Even way back in 2005 the company was forced to concede over $100 million in services discounts in North America alone. The margins lost along with that $100 million lowered the entire company's annual profitability by 10%. Thanks to implementing a series of price defense recommendations from TSIA, along with some heroic improvements in service delivery, the company got a grip on this problem. But we know of enterprise software companies whose largest maintenance customers are demanding—and receiving—discounts of 30% to 40%.

MAINTENANCE MATH

For all these reasons—an eroding value prop, declining product prices pulling down the corresponding service prices, and increasingly sophisticated negotiating techniques by customers—top-line revenue from service, support, and maintenance businesses is under increasing pressure. The best you can do with an offering in the "Decline Stage" is to tightly understand the elasticity of demand for each product and customer segment. To be as smart as you can about pricing and packaging for as long as you can.

The cold, hard fact is that average service revenue per unit in the field is dropping. If you're a hardware company, you might measure it in physical units; for a software company, it might be per user license. Either way the trend is down and it's getting worse. This is a critical fact not talked about nearly enough by most management teams.

This is because some companies have found a way to mask these problems—but not to make them disappear. Over the past few years, these companies' total maintenance and support revenue has actually increased, despite the drop in service revenue per unit. How is that possible? The trick is to add enough new maintenance customers to offset the reduction in per-unit maintenance revenue. Some do this through successful selling; others do it through acquisition—essentially buying a company for its customer service business. Oracle has taken this practice to a whole new level. Why do you think they would pay billions for PeopleSoft, BEA, and a dozen more "mature" software companies?

But there are some significant underlying problems to this approach. To see what we're talking about, let's go through a very simple example.

Let's say you have a thousand systems in the field, each one generating an average of $10 thousand in maintenance and support annually, for a total annual revenue stream of $10 million. Now let's suppose:

- That you have a 90% contract renewal rate but your product sales force is able to add 100 new product purchases per year so that your total number of active maintenance contacts stays flat.

- Next let's assume that you're selling those newer products for less than the ones they are replacing (which is exactly what happens in the market). The result is a 5% per year decline in your average contract value.

- Finally, let's assume that your customers are able to achieve a 10% average discount rate on renewals. Here is what your customer service revenue top line will look like over the next five years:

	Baseline	Year 1	Year 2	Year 3	Year 4	Year 5
Less 10% Attrition		900	900	900	900	900
Add 100 New Units		100	100	100	100	100
Units in the Field	1000	1000	1000	1000	1000	1000
Service Revenue per Unit	$ 10,000	$ 9,500	$ 9,025	$ 8,574	$ 8,145	$ 7,738
Total Gross Service Revenue	$ 10,000,000	$ 9,500,000	$ 9,025,000	$ 8,573,750	$ 8,145,063	$ 7,737,809
Minus 10% Discount	$ (1,000,000)	$ (950,000)	$ (902,000)	$ (857,375)	$ (814,506)	$ (773,781)
Recognized Service Revenue	**$ 9,000,000**	**$ 8,550,000**	**$ 8,122,500**	**$ 7,716,375**	**$ 7,330,556**	**$ 6,964,028**
% YOY Growth		-5%	-5%	-5%	-5%	-5%
Cumulative Growth		-15%	-19%	-23%	-27%	-30%

Next let's suppose that you have a 90% contract renewal rate but your product sales force is only able to add 75 new product purchases per year so that your total number of active maintenance contracts declines. Keep the rest of the assumptions the same as above. Here is what your customer support and maintenance revenue top line will look like over the next five years.

	Baseline	Year 1	Year 2	Year 3	Year 4	Year 5
Less 10% Attrition		900	877.5	857.25	839.025	822.6225
Add 75 New Units		75	75	75	75	75
Units in the Field	1000	975	952.5	932.25	914.025	897.6225
Service Revenue per Unit	$ 10,000	$ 9,500	$ 9,025	$ 8,574	$ 8,145	$ 7,738
Total Gross Service Revenue	$ 10,000,000	$ 9,262,500	$ 8,596,313	$ 7,992,878	$ 7,444,791	$ 6,945,632
Minus 10% Discount	$ (1,000,000)	$ (926,250)	$ (859,631)	$ (799,288)	$ (744,479)	$ (694,563)
Recognized Service Revenue	**$ 9,000,000**	**$ 8,336,250**	**$ 7,736,681**	**$ 7,193,591**	**$ 6,700,312**	**$ 6,251,069**
% YOY Growth		-7%	-7%	-7%	-7%	-7%
Cumulative Growth		-17%	-23%	-28%	-33%	-37%

This is not a pretty picture. Remember that your overall service revenues can go down for any of these reasons. It could be that total discounts begin to exceed new contract revenue. Or it could be declines in service prices exceeding new contract revenue. Contract cancellations are not the only demon to fear. As these trends accelerate, growing top-line customer service revenues will become more and more difficult. And with a large—and growing—percentage of total revenue coming from services these days, it could become nearly impossible for affected companies to grow their overall business simply because product revenues are flat and services revenues are dropping faster than the product sales team is adding new service contracts to cover the contraction. The tough economy of 2009 is already proving this to be true as some major companies are reporting shrinkage in maintenance revenue.

To add insult to injury, the final threat to the classically profitable maintenance and service business is that the cost of providing those services is likely to increase from this point forward. Over the past decade or so, many tech companies have done a marvelous job of improving their service economics by applying a combination of manufacturing process principles, technology, and global labor arbitrage. But now they are facing declining returns from all three.

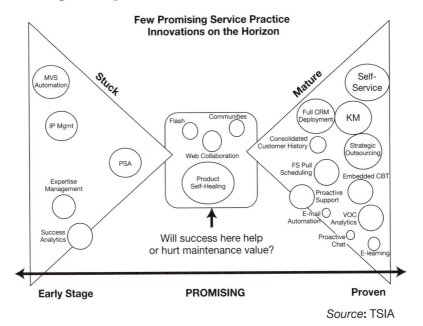

Source: TSIA

FIGURE 3.1 Maturity of Service Tools.

Many companies are as outsourced as their customers will put up with. And the cost of labor in most typical outsourcing destinations is growing much faster than the cost of labor in the United States, thus providing less attractive savings each year. Unfortunately, we will not be able to raise maintenance prices to keep pace. The result will be lower margins.

As for counting on further service efficiency gains, based on TSIA's in-depth review of the technology available to service organizations today, most of the big wins might be behind us. Innovations in technology for services delivery—including customer adoption of self-service tools—have plateaued. For example, knowledge management—often seen as the enabler of tremendous gains in efficiency—seems to have hit a wall. According to TSIA data, end customers are averaging just a 39.6% success rate using Web self-service in 2008/2009—a number that has not improved appreciably in the last four years.[2] So if your company has been aggressively investing in service automation and you are already using all the leading categories of tools, you might already be seeing an uncomfortable future for your operating economics. Can we design even more efficient service processes and systems in the future? Absolutely. However, even taken together, the efficiency gains may not keep up with shrinking service revenues and might soon produce a serious, large-scale economic challenge to the primary profit engine of many tech companies.

An important point needs to be made about the middle box in our chart. The largest efficiency opportunities that remain for customer service may actually be ones that are not in a tech company's economic self-interest. Using technology to prevent product problems, doing more remote solves, improving product reliability and offloading basic service issues to your user community is clearly good for your customer (and your costs!) but may reduce your perceived service value to them at the same time. It also makes it harder to differentiate your service from that of your competition. These challenges provide yet more reasons why we need to "jack up" the customer maintenance/customer service value prop and replace the engine ASAP.

Many a CFO has rebelled at the idea of doing anything that might affect the revenue or profits of this service line. "Don't Mess with Maintenance" could be a popular T-shirt for tech company financial execs. At TSIA we believe the bigger risk over the next few years is NOT to mess with maintenance. It is no longer a safe assumption that the maintenance business will continue to grow and become more profitable with each passing quarter. And making the changes needed to preserve it will take years, not quarters.

CONSUMERS HUNGRY FOR NEW KINDS OF HELP

We have been spending quite a bit of time on enterprise technology markets. While most of the phenomena (complexity avalanche, consumption gap, etc.) apply to virtually every tech market, the quirky consumer side of tech services is a different story with a different problem: a revenue problem.

There are two specific historical events that we believe contributed mightily to consumers' resistance to paying for technology services. On the hardware side, it was the ridiculous battles of the mid-1990s, waged by Compaq and other PC manufacturers who hoped to increase their market share by offering customers longer and longer warranties. Remember the 90-day warranty? Compaq pushed that to three years, and everyone else had to follow suit. Since those warranties were longer than the useful life of the product, consumers didn't need to buy extended warranties, so for a long time, almost no one tried to sell any. It has taken over a decade for today's PC manufacturers to scale back to a reasonable level of warranty coverage—and to discover that money can be made on higher levels of service. The good news is that their extended warranty revenues are growing at a faster rate than ever. And there's often a nice margin on the parts business. But considering the cost of providing the service, the margins are often pretty small. Most consumer companies actually make an accounting move that shifts revenue dollars from the sale of products to the services revenue category to compensate the services organization for costs during the warranty period. But we would argue that most consumers don't see any of their purchase price as going toward that services warranty value proposition. So, except for some of the large consumer tech companies who have very profitable service businesses, much of the cost of delivering consumer service still has to be subsidized by product margins. That is a tall request when competitive forces are driving hardware margins in categories like PCs, TVs, cell phones, and the like down to single digits.

Software support in the consumer space has its own slightly absurd story. In the beginning (late 1980s and early 1990s) there were products like WordPerfect, Lotus 1-2-3, Quattro Pro, and Harvard Graphics. These products sold as individual modules for prices like $395, $495, and $595. To buy a top-rated word processing program, along with a spreadsheet, a database, and a presentation package, would cost you about $2,000. At those prices, PC software companies didn't mind taking a phone call or two to help out buyers who needed some kind of support. Many of these companies had customer support organizations that simply emerged, not to achieve competitive advantage or as a quest for profits, but simply because the phone was ringing.

Then competition and consolidation came to the PC software market. Microsoft started bundling more applications at a lower cost, and everyone's prices came down. Support costs as a percentage of revenue began to climb. Some companies reacted with a strategy best known as "bad support." It is a fact that consumers have been far less satisfied with support services than commercial or enterprise customers. I got to witness that personally from a front-row seat. For many years I was CEO of a company called Prognostics, which was the leading provider of customer satisfaction surveys to high-tech companies in the 1990s. We did the survey work for nearly 200 leading firms spanning most every sector of tech. We watched in awe as many a consumer tech executive stood by and allowed horrendous sat scores with their customer service to exist. They couldn't afford (or wouldn't invest) to fix the problem. It wasn't that the solutions were unknown: They were being executed every day by tech companies in the enterprise space. There were even companies who served both enterprise and consumer customers, and they all had much lower sat scores in their consumer segments. Why the big difference? It's simple. One customer segment (enterprise) was paying for the service and one segment (consumer) was not. The result for consumers was long hold times and slow answers. Compared to enterprise support, it was unsophisticated and frustrating to use. Vendors did everything they could to keep the customers from calling. They hid their 800 numbers, limited their hours, and even made the consumer pay the long-distance tolls.

It's also why companies initially raced to low-cost offshore outsourcers to provide their service.

I will never forget when Microsoft made the bold and brave move in the mid-1990s of actually trying to charge for PC software support in an aggressive way. Companies like WordPerfect refused to follow suit. I loved Microsoft for trying, but after a major revolt from the press and their customers, it basically gave up on a high-profile marketing approach.

So from the very beginning of personal digital technology, consumers and tech manufacturers developed a tacit agreement: "You give us crappy service, and we won't pay you for it." It was and partly still is the part of the industry that the media and consumers love to hate. We once had a discussion with an editor from a major U.S. newspaper about tech services. He told us, "The only good story about support is a story about bad support." This attitude by the press didn't do much to improve the reputation of consumer tech support. But consumers bought products anyway, and bad service rarely cratered a hot product. Most senior consumer tech execs sat back and allowed the problem to continue.

While consumer tech companies today are doing much better than at any previous time in history at charging for their services and improving their quality of service delivery, the concept of making consumers pay for extended and/or traditional service contracts is still a pretty tough sell. Only the consumer companies with serious scale are making money at basic service and support.

But once again, the complexity avalanche threatens to shake up the status quo. Where is this sea change coming from? Ten years from now we'll look back and see that it all started with home theaters. Unlike basic stereo equipment and stand-alone TVs and PCs, setting up a home theater is nearly impossible for the average consumer to comprehend. There are so many options, so much complexity, and so many wires. You might as well ask people to build their own cars. As a result, thousands of small "home automation dealers" around the world began to design, sell, install, and service home theater systems. The ground these people broke was important. They were the first ones with the guts to show their rich consumer clients that part of what they were paying for was service—not so much extended warranties, but billable time to install this complex system in their home. The complexity avalanche had finally become scary enough that well-heeled consumers gave up and stopped trying to do it themselves. And for maybe the first time in modern consumer technology history, customers were glad to pay up. Then the Virtuous Cycle started. These service providers became the trusted adviser to their clients and they began to pull through products. Now they are a major sales channel for high-end home video and audio manufacturers.

Also today, services have become the fastest growing part of consumer technology sales at big retail stores like Best Buy. And it's not only home theaters. Companies are now selling services to install home computer networks, lighting controls, and surveillance systems. And most important of all? Apple is selling access to a "Genius." This may be the second major milestone in the emerging world of fee-based consumer services. What's so significant about the Apple Genius Bar? It's quite simple: Apple is providing basic assistance—*very* basic—for a fee, and people are happily paying for it. Just as home theater buyers paid for services to bridge a major complexity gap, Apple customers are now paying a fee just to get access to someone who can answer their everyday questions—the kinds of questions that they hope will save them time and frustration and get the most value out of their products.

Now, just about every retail store has its resident Geeks and Gurus and Geniuses. It's gotten to the point that they're almost "cool" heroes to their customers. Are we on the brink of a new tacit agreement between consumers and tech companies where consumers actually pay for services and get great service in return? Could the ugly duckling turn into a sexy goose? We think so. Just keep your eye on the rise of services in consumer tech over the next decade!

So whether it's enterprise and commercial-class maintenance services that are having their value propositions challenged, or consumer markets awakening to the true value of good service, the message to manufacturers and software companies is clear: Support services can no longer be a low-level focus for executives, an afterthought to their product business. Those who invest to get it right will improve their bottom lines; those who don't will suffer.

PROFESSIONAL SERVICES: HIGH IN DEMAND, LOW IN PROFITS

Another popular set of technology services purchased mainly by enterprise, commercial, and government customers is commonly called professional services (PS). This is where the vendor sends out teams of people on a project basis (usually consulting engineers of one kind or another) to their customers' sites to install the new product, integrate it with other existing applications or products, migrate legacy data from the old to the new, and maybe even redesign a few business processes.

Due to the complexity avalanche, these are services that customers want and need. Demand is growing rapidly. Unfortunately, flying highly trained people all over the world and having them live out of hotels for months at a time makes professional services extremely expensive and produces minuscule (sometimes negative) margins. In addition, thanks in part to SaaS, more enterprise customers are bristling at having to pay for a complicated deployment in the first place—and that puts even more pressure on professional services margins. There is at least one actual case of a large software company whose professional services organization had operating losses that were equal to their revenue—both were in the hundreds of millions of dollars!

Margins can be so low in professional services that one large tech CEO (only half-jokingly) told his senior professional services executive that if he could, he'd close up that entire part of the business. This is easy to understand when you think that software gross margins are in the 80% range and professional services margins are more like 10% to 30%. Operating profits are often +/- 0.

Concerned about the dilutive effect that these low margin revenues have on the company's overall gross margin percentage, more than a few software companies are putting an artificial cap on how much professional services revenue will be allowed to grow. Here's how that might work at a hypothetical software company and why management might see it as the right move to keep their stock price up.

Let's say a company brings in 40% of its revenues from product sales, 40% from customer service and maintenance, and 20% from professional services. And let's say margins are 90% on software, 80% on customer service, and 10% on professional services. This would create a blended, overall gross margin for the company of 72%. If revenues from professional services were to increase from 20% to 30% of the total (which would mean the percentage from sales and customer service would each drop to 35%), gross margins for the whole company would slide by 6.5%, down to 65.5% for the period. This is not something that Wall Street typically responds well to. By capping your professional services revenues, you actually prevent a margin percentage dilution that occurs simply by a shifting mix of revenue. Or so the theory goes. But let's attack this idea of professional services being dilutive to a software company's margins. The question is what to prioritize—gross profit margins or gross profit dollars? If your answer is the former, then you will cap the growth of your professional services business. If it is the latter, even a marginally profitable professional services business is a good move. It makes some contribution to EPS, but more importantly, it helps customers deal with the complexity avalanche, and that will drag more product revenue with its corresponding services.

If you are a typical hardware company, you may view the gross margins from professional services in a completely different light. Many competitive hardware product markets have to scrape and claw for positive product margins. Between competitive pressures, the cost of marketing and sales, and incentive payments to resellers, there simply isn't much money left. To these companies, making 20 points of margin from their professional services portfolio sounds darn good. This is why you will find hardware companies like IBM or HP much more interested in growing their professional services business than software companies like Oracle and Microsoft.

The other critical players in the enterprise professional services market are the "pure service" providers like Accenture, EDS (now a division of HP), and their offshore counterparts like Wipro and Tata. These are companies whose *only* business is providing a variety of technology services to large enterprise or government customers. For these companies, service margins are their *entire* margins, so they tend to be very disciplined and watchful over their economics. From this group of pure service companies has emerged a set of margin myths that remain the bane of professional service executives within product companies. Thomas Lah, the executive director of TSIA, talks about the 40%/20%

margin myth of professional services: "For some reason, product company CXOs feel their embedded professional services firms should be achieving at least a 40% project gross margin and 20 points of operating margin. This, historically, was a very common business model for top system integrators. Unfortunately, very few firms today beyond the offshore providers are able to achieve these targets." Until TSIA, embedded professional service executives simply lacked the data to prove what the valid comparison really was. Many tech product company CEOs and CFOs were free to look at the profitability of businesses like Accenture and offshore service providers and then conclude—sometimes incorrectly—that their own professional services organization's financial performance is sub-par. This is just one reason why the tenure of professional services execs in the IT industry is probably the shortest of any other line executive. As a rule of thumb at TSIA, we expect senior professional services executives to last an average of just 18 months in their jobs.

Want the truth about margins in the professional services business? According to data from the TSIA Service 50 and our professional services industry benchmarks, the realities of the professional services and technology consulting business look like this:

- At the beginning of 2009, average gross margins of pure global service providers such as Accenture and Infosys began around 17% and went up to the mid-40s.[3]

- Average gross margins of professional services organizations embedded inside product companies experience a wider range of results. Margins for these embedded businesses at the beginning of 2009 began at +/- 0% and again topped out in the 40s.[4]

What's the difference? Why could "embedded" service businesses be inherently less profitable than independent service businesses? To overly simplify the economic model, one could say that professional services profits are a function of five key variables:

1. Ability to "get your price."

2. Cost of labor.

3. Efficiency of process and technology.

4. Ability to capture and reuse IP.

5. Of course, your ability to execute.

An objective business analyst *could* assert that professional services uses an inherently inefficient productivity model focused too much on labor and has an underdeveloped process for capturing and reusing the intellectual property that is created during customer projects. That argument certainly could be made. The labor component is obviously a main reason why the offshore service providers have higher overall margins than many of their western counterparts, despite the frantic race by many U.S. companies to get more of their labor cost structure moved to lower-priced markets like India. But offshore or no, it is a tough nut to crack. While most pure service companies seem to be better at these other practices than embedded service businesses, neither have a stellar track record to date. What everyone can agree on is that moving professional services to the Age of Tools is an area ripe with improvement opportunities over the next decade. IBM has initiated a massive project to automate and streamline traditional professional services to make them more like Web services in what could be a watershed initiative that brings entirely new levels of profitability to this business. In addition, Dr. John Ricketts, a consulting partner and technical executive in IBM Global Services—and an IBM Distinguished Engineer (IBM's top technical honor)—has just written a fascinating book, *Reaching the Goal: How Managers Improve a Services Business Using Goldratt's Theory of Constraints*. In his book he talks about how IBM reapplied this industrial management technique to the complex, diverse, and unique business of technology professional services. The goal, as Erik Bush, VP IBM Global Services, says, was about "relearning the services business from the ground up, with a focus on getting to the truth of genuine cause-and-effect relationships between the elements of our (service) business and the management systems we deployed to drive its improvement."[5] This modeling and understanding of the services business is an essential step toward putting technology tools in place that can boost revenue and improve profits.

But the main reason why embedded service businesses historically failed to meet the unrealistic margin expectations of CEOs and CFOs was typically their inability to "get their price." When customers bought a product, they wanted a business result. The fact that your product required millions of dollars in ancillary services spend to become useful was not considered a positive by most enterprise buyers. They could understand that a pure service company had to make a profit. But you? They knew that you badly wanted the product sale. They also knew that your product's complexity was what was causing them to spend all these service dollars in the first place. So how guilty did they feel about negotiating the living hell out of your services proposal or your bundled solution? Not very. And since product companies historically preferred to allocate revenue from the "bundled price" more into the product category and less to the services category, the professional services P&L always looked bad.

Ironically, the professional services P&L has been getting help from an unlikely source: the U.S. government. Revenue accounting rules like VSOE (vendor-specific objective evidence), driven by FASB (the Financial Accounting Standards Board) and the SEC (Securities and Exchange Commission), has led to specific regulations and guidelines that tech companies must use to categorize and recognize revenues when selling customer "bundles" that include both products and services. VSOE rules are driving companies to limit the discounts on service to within a fixed range so they can defend the market value of their services with auditors. If a product company cannot defend a consistent price for its professional services, the whole "bundle" might be classed as a service, and the company couldn't recognize product revenues until the services are completely delivered. That defers badly needed product revenue from this quarter to some future quarter. So the days of wildly fluctuating discounts where sales reps "threw in" the professional services to get the product sale are behind us. This is bringing up professional service revenue to better reflect its true value. This may make the gross margin comparison to the pure service companies a fairer argument.

The bottom line is that most professional service offerings are still labor- and travel-intensive activities that are far less profitable than either selling software or customer support/maintenance contracts. Software companies offer them because they have to, and systems companies do because they want to.

But don't forget, the need for these advanced services is growing. They help overcome part of the complexity avalanche that customers encounter on the road to product value, and they are in increasingly high demand. Rather than being a growing profit drag for some companies, the challenge over the next decade will be to systematically improve the profitability of the professional services business by applying many of the same tactics already successfully in use by the customer service function. Will that be more difficult in professional services? Yes. Can it be done? We believe so.

4 | The Value Added Service Model

At the Technology Services Industry Association we believe that our industry needs to rethink the way our service strategy, organizational structure, and investments map to our customers' needs.

Here's our core thesis in a nutshell:

- Technology companies are already suffering from a growing consumption gap caused by the complexity avalanche they have unleashed on their customers. This gap is negatively affecting the sales of products and services across most every industry sector.

- Developing an effective service approach to the problem could be a good stand–alone business, and would drive more frequent and larger product repurchases along with a host of other financial benefits.

- Today's existing professional services function has the right goal of moving customers successfully along the product adoption life cycle. It also has the right skill sets. But it stops short of end-user adoption (EUA) and is a very expensive financial model.

- Today's customer services function is exactly the right operating model but has the wrong customer mission and lacks many of the right skill sets.

What the technology industry needs is a way to deliver the next generation of adoption services using the customer service delivery model.

But in order to do that, the industry will have to say goodbye to business as usual and hello to a completely new way of thinking. This is a watershed moment in the world of technical services: the end of the *availability service era* and the birth of the era of *value added service* (VAS). Here's the difference:

	Availability Era	Value Added Service Era
Mission	The primary purpose of service is to establish, maintain, and optimize system availability.	The primary purpose of service is to proactively accelerate the consumption of product value and lower the cost of ownership for every customer.
Advantages to the Vendor	– Predictable annuity revenue stream – High gross margins	– Predictable annuity revenue stream – High gross margins and a better value prop to defend them – Reduced cycle time between customer repurchases – True competitive differentiation and market advantage – Higher product margins – New levels of customer loyalty and customer satisfaction

The best news of all is that the delivery asset for VAS is mostly bought and paid for. Yes, it will need lots of retooling, but the "factory" is there. If you add up the best parts of each service silo, the systems and the management for VAS are already in place. It just needs the pieces reorganized and then given a new mission.

If you take a look at the following chart by Neochange, which illustrates the way tech companies typically spend their money today, you'll begin to appreciate how far away we actually are from confronting this problem.

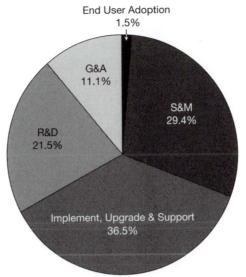

Source: Neochange analysis of software company expenditures

FIGURE 4.1 Providers Are Currently Underinvesting in Activities that Enable Customer Success.

- Research and Development – 21.5% of the budget.
- Sales and Marketing – 29.4% of the budget.
- G&A – 11.1% of the budget.
- Traditional Professional and Customer Services – 36.5% of the budget.
- End-User Adoption Services – 1.5% of the budget.[1]

If you were to give your customers the power to allocate your budget, do you think they'd follow this model? Not even close. You can bet they'd have you spending more than 1% of your budget on end-user adoption. As we've discussed throughout this book, end-user adoption is the path to value for your customers. And for your investors, it's a path to increased share price.

How many times have you heard a frustrated customer tell a member of your staff, "You guys *have* to know how to do this!" And you know what? They're right—you *do* know how to give them the value they want. Depending on the type of technology product you sell, your sales and service staffs could have a hand in hundreds or even millions of new product deployments each year. Whether they realize it or not, they're learning what customers want to do with the products they've bought and how they go about doing it. They see successful practices and unsuccessful ones.

So where does all that learning go? For most companies, too much of it resides in the minds of individual employees. For the customer service reps, taking the time to document that kind of usage information is outside their charter. Professional services engineers often keep the knowledge to themselves because of the "knowledge is stature" phenomenon. Only a small percentage of that deployment learning is captured, managed as a corporate asset, and redistributed to other customers. Every stakeholder in your company—whether a customer, shareholder, or employee—is paying a price for this lapse. A growing number of enlightened executives are agreeing that this must be addressed. The question is how.

VALUE ADDED SERVICE: A NEW LEVER FOR MANAGEMENT

Let's look at a basic product adoption life cycle model.

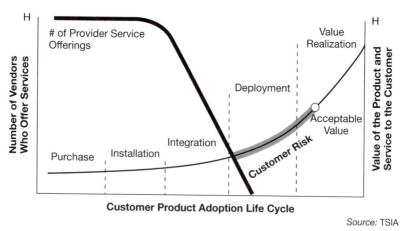

FIGURE 4.2 Value Added Service: Seeing the Consumption Gap.

On the day a customer buys a product, he or she has some expectation of realizing value. If the customer is a large hospital chain or bank, it may have already received an extensive, written ROI analysis from your company identifying the specific cost savings or business value that your product will give it and a proposed time frame in which to realize it. If the customer is a consumer who received a digital camera for Christmas, he or she may be mentally willing to spend 20 minutes learning how it works so he or she can take pictures of the family dinner that night. Anything more than that and the customer will become frustrated.

Overall, whether it takes 20 minutes or 20 months, every tech user in the world has an explicit or tacit expectation of acceptable value and how long it's going to take to realize it. In most cases, value can be achieved only once the product's features are in use. Unfortunately, as reflected in our model, that's the very last phase of deployment.

In the first stages, as we discussed, vendors provide an impressive array of sales, installation, and implementation services either directly or through their partner network. The need for and scope of these services vary greatly by product category. A digital TV will require different services than an iPod or the global deployment of a secure data network. But generally, within any specific product category, almost every provider offers a similar set of services.

The craziest part of the model is how our service offerings align (or don't) with the realization of product value by our customers. As we move to the right on our chart, the value to the customer *increases*—and the number of service offerings *decreases*. By the time you get to the value realization phase—where individual end users are attempting to exploit the features and derive personal or business benefit—our service offerings are almost nonexistent. In the case of enterprise technology, we pulled our staff out once the professional service and education parts of the deal were fulfilled. Both of these phases generally take place *before* customers have actually begun using the product live. Once they do, we simply offer services to keep the product available (maintenance) with no specific concern for how much or how little value is being realized by the customer. Amazingly, during the value realization phase of the adoption cycle where the product's true value to the customer will be determined, we have no services available to ensure the outcome.

As a result, there's a risk period that begins with your organization's last proactive customer contact and lasts until the customer actually realizes your product's value. This is a critical time. Here is when customer outcomes are won and lost, when they achieve success or failure. Most of the time, though, it's something in between. (Remember from Chapter 1 the 74% of enterprise software customers who achieve only "moderate" success.) We've all had some personal experience with this problem. Somewhere in your home you, too, probably have a drawer full of consumer technology failures—products you were never able to get working, products where the "customer as general contractor" strategy (you in this case) failed. We're guessing that for most of the unused products stuck in your drawer, you won't buy the next generation of these same products that are already gathering dust unless there is a mighty compelling reason.

As a tech manufacturer, you may think that we are barking up the wrong tree—that what really needs to happen is that you simply need to make your next-generation product easier and more compatible. Won't that reduce the consumption gap? Won't that get the consumer to buy from us again without our having to provide any more services? Well, you could be right. There are some examples of products that finally got it right and made a market out of a promising technical capability. But it is a *very* low percentage of the total new products that are released each year. The rest are going to need help devising a strategy to deal with their version of the consumption gap.

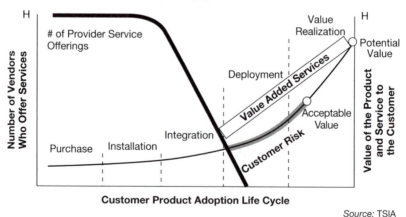

FIGURE 4.3 Value Added Service: Serving the Consumption Gap.

We're quite sure no one would disagree that the tech industry has a vested interest in ensuring that we deliver the value that our customers expect. The trouble is that we're not proactively managing value realization. That's a shame—and we need to fix it. But even that isn't the primary goal of VAS. The truly big win is not in moving your customers to a level of acceptable value but to stay with them until they realize what's *possible* with your technology—delivering not just the minimum acceptable value but the maximum. That's what VAS is all about—and it's a key way that vendors of the future will compete and differentiate. Think of the ROI to customers! In the case of corporate customers (enterprise or SMB) they have spent huge amounts of money just to get to the Value Realization phase. They bought your product and any other products or services that are needed to make your product work. And so far, all this huge commitment of time and expense has just gotten them is to the brink of actual value realization. What if for 20% more they could not only ensure a solid value outcome, they could hit (what in baseball we call) a home run?

What exactly does that mean? Let's take a look at a hypothetical B2B example.

Let's say that Barclays Bank purchases SAP CRM. SAP and its partner Accenture are hired to help Barclays through the usual purchase, installation, and integration phases. The team develops a strategy for deployment that is supported by SAP's end-user education and some project management from Accenture.

The go-live date comes and goes with no major crises and users are beginning to use certain parts of the product according to the plan. Just like teaching a child to ride a bicycle, SAP and Accenture begin to pull away. That is, after all, what the project plan called for—and what Barclays paid for. SAP and Accenture have delivered a functioning system that the customer is using. Everyone would consider the engagement to be a success.

Now Barclays is on its own—wobbling, nearly falling, but moving forward. Have its users achieved the value called for in the ROI analysis that SAP and Accenture provided? Not yet; they're just on the road to get there. Along that road any number of things could and probably will happen. It's now up to Barclays to fully deploy the technology and to assume the entire risk of delivering the business benefits it sought (and thought it had paid for). Based on our research the odds of success aren't very good:

- 1 in 1.4 chance that Barclays will be only moderately successful.
- 1 in 8 chance that Barclays will be unsuccessful.
- 1 in 7 chance that Barclays will be very successful.

These are pretty dismal odds—and that's just to get Barclays to a level of value it deems acceptable. But there's an even larger win at stake here, one that Barclays probably doesn't even fully understand. You see, Barclays is not the only bank running SAP.[2] The Bank of Ireland is, and so are China Everbright Bank, Europe's Commerzbank, UBS, and dozens of others. Each of these banks is doing some things uniquely well, finding specific applications that provide very high rates of return on their usage of the product. What if SAP was capturing all those best practices? If SAP could take the best functions from each of those banks and make those processes, tools, and information available to the rest of the customers' staffs, it would have a banking application that wouldn't be simply acceptable—it would be superlative.

Actually, SAP already has most of that information. Where is it? And why isn't it finding its way to the individual users and IT team at Barclays? We're not talking about just giving them a white paper, but person-by-person, team-by-team guidance on how to best realize value from their investment—in a word: VAS. The good news is that most of the resources required to do exactly that are already in place—not only at SAP, but at most tech companies as well.

So let's give it another shot at Barclays, this time employing a VAS model during the value realization phase.

This scenario starts off just like the first one, and everything works the same way through the deployment phase. Again, the go-live date has come and gone, the SAP and Accenture professional services teams are pulling out. End users are now beginning to use the product. The only thing that's different is that SAP's software is now tracking which users are using which specific features. This information is being aggregated and sent to SAP's VAS organization at one of its customer support centers. There an application is mapping Barclay's use of the product and comparing it to a best practices model that SAP has aggregated from all the banks that currently use its products. The application is looking for gaps—in planned deployment versus actual deployment, in actual deployment versus best practice.

Now because SAP is tracking activity at the user/feature level, the application is able to identify the features that users need the most help using. Armed with this information, an SAP VAS representative sets up a meeting with the Barclays deployment team. The representative explains that Barclay's marketing arm is not really using the part of the SAP product that analyzes real-time offer management. The best banks that use SAP are realizing a 30% better click-through and adoption rate on their promotions than Barclays. Barclays agrees to have SAP proactively address these usage gaps. SAP may do this for a fee, based on a new VAS contract, or as an entirely new service offering contained within the annual support and maintenance agreement that Barclay's buys from SAP each year.

THAT is customer service—the kind that will differentiate companies going forward, the kind that turns the complexity avalanche from customer frustration to customer success.

OK. We know that you all want to deliver this result for the customer. The big question at this point is how can SAP (or your company) economically deliver all this high-value content? Here's where the economic beauty of VAS becomes clear. Unlike the end-user tracking capability that most tech products don't have today (and which will need to be created), the actual delivery assets of VAS are, for the most part, already bought and paid for.

TODAY'S CUSTOMER SERVICE MODEL: GREAT ASSET, WRONG MISSION

The primary organizations within most tech companies can be plotted out as in the following graphic:

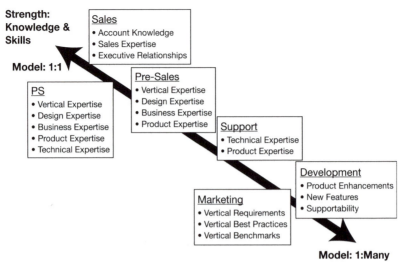

Source: TSIA

FIGURE 4.4 How Tech Companies Organize to Deliver Value.

At one end of the spectrum are the organizations inside your company that are built to deliver value to customers on a one-to-one basis. These include your sales, sales support/pre-sales, and professional services teams—highly skilled, highly paid individuals with great customer-facing attributes. You send them wherever they're needed to make specific, critical-deal milestones happen. Or you might be getting some or all of this done through a network of partners who receive an attractive product discount structure in return. Either way, it's an expensive model.

At the other end of the spectrum are the organizations that are built on a model and infrastructure designed to deliver value to customers on a one-to-many basis. The ultimate one-to-many organization is your product development team. They build a product and exact copies of that product get shipped to customers around the world. One of the best examples of this concept is SaaS. A change to a product can be readied at 12:01 a.m. and every SaaS customer has access to that change at 12:02 a.m.

Another one-to-many organization is marketing. The marketing team designs offers, messages, packaging, or whatever is needed to make the job of selling your products more efficient. Once they get it, they replicate it and disseminate it as widely as possible—through your sales force, Web sites, advertising, white papers, and so on.

But as nice as development and marketing's scalability are, they're missing one key element. They are not customer-facing organizations. They do a great job, but there's only one customer-facing organization that was built from the ground up to be one-to-many, and that's customer support (or customer service, tech support, or whatever you call it). This organization can safely claim to be the most cost-efficient, customer-facing organization in the company. Some firms may have a field service organization as well as a centralized, customer-support function. In the case of software-only companies, it's really only the latter.

Here are nine reasons why today's state-of-the-art customer support organizations are the ideal platform upon which to build your VAS delivery model:

1. **Key call centers are already in locations around the world where labor and telecom costs are as low as possible.** Whether they are badged employees, outsourcers, or a combination of the two, most technical support or customer support organizations have made major investments in either offshore delivery or moved to low-cost markets in North America.

2. **All the centers are tied together into a seamless global network.** This allows remote teams to work together on a customer's problem either in parallel or in serial, working the case around the clock or teaming up to answer questions with incredible coordination and efficiency.

3. **They are among the best CRM users in the company.** Customer service organizations often use a common technology platform that includes modules of enterprise-wide software applications that allow information to flow freely between service, sales, accounting, manufacturing, or any other part of the company. They are adept at operating in a highly efficient, systems-driven environment where customer information is routinely and effectively captured. In general, customer service is the best measured of our customer-facing functions.

4. **At the heart of the customer support organization is a commitment to formalized, effective knowledge management.** This means that the people, practices, processes, and tools are in place to ensure that once a question has been answered anywhere in the world, the answer can instantly be accessed and utilized by every service employee in the company or every customer in the world. There is no need to recreate the work that went into answering the question the first time when another customer calls a different agent with a similar problem.

This simple process is a vital and unique attribute of the customer service and support organization. Almost all the other customer-facing organizations in the company are staffed with employees whose personal knowledge is directly related to their status and income. A consultant in the professional services organization is billable at a higher rate (and earns more) if he or she knows more than a colleague. In sales the situation is similar. A more knowledgeable salesperson can usually make more money and get more recognition than other, less-expert peers. As a result, neither sales nor professional services have made much of a commitment to (or registered much real success in) managing knowledge effectively as a means of reducing costs and improving responsiveness.

But in customer service, knowledge management (KM) practices and technology are the bedrock of the business. Not only do employees learn practices that optimize KM, but they earn compensation and recognition as rewards for their contributions to the organization's knowledge base. Employees are rewarded not just for *knowing* more but for *sharing* more. The more entries a person makes into the KM system, the greater the reward and recognition. That is exactly the behavior management wants. This orientation is far stronger in customer service organizations than in their organizational counterparts within the company, and will prove vital as we move toward VAS.

5. **Their technologies already extend to and are used directly by the customer.** The commitment to technology that you find in most customer service organizations doesn't stop with employee-usable tools. These same organizations typically own the most

sophisticated customer-accessible technology platform aside from the product itself. Customers can access the KM system via the company's Web site to answer their own questions, check the status of open issues, and even schedule a field service visit at a time that works best for them. It's a remarkably robust set of tools that every customer can use.

6. **They already have access permissions to most customers' systems.** In most tech companies today, the customer service organization routinely accesses customers' systems. In the enterprise world, that means they have the security permissions and business processes that allow them to see how the product is being used. In consumer product markets, customer service agents routinely take control of any device connected to the Internet via remote control applications. Using this technology, a support person sitting in a cubicle in Bangalore can see what a Chicago-based customer has done and can lead him through a tricky process. These organizations often work with the development teams to design, distribute, and use automated "phone-home" technologies built into the product. Since this is one of the key components of a successful VAS strategy, experience with existing diagnostics tools makes them an ideal destination for EUA information captured by the product. Sure, some government or other markets will be more sensitive to the security of their user data than others, but for many customers it's a small price to pay.

7. **They are pros at the soft skills side of working with customers.** Most customer service organizations have formal training programs, certifications, and measurement systems designed to ensure cooperative, satisfactory communication directly with end users or IT staff. No other organization in the company is typically as invested in—or as sophisticated at—consistently delivering an excellent customer experience from anywhere in the world.

8. **The field service organization has unique physical access to the customer.** Field service reps have visibility into not only your products at the customer site but also the work flow of the customer around your products. They see competitive products and how they're being used. Many of these reps have earned enough

trust and respect from the customer to suggest change. Since most field service organizations are trying to find new ways to make their reps more billable and more contributory to the bottom line, a role in the VAS strategy could be just the ticket.

9. **They have the lowest-cost labor of any customer-facing organization.** While this may seem redundant to our first point, it actually is not. This is not about where the labor is based but about the relative income levels of the labor. Compared to sales, consulting, or professional services, ordinary customer service employees are a real bargain. We have established university relationships and multilingual hiring strategies that result in a steady stream of low-cost, capable talent. We are learning more about how to take advantage of high-quality, lower-cost work forces like retirees, the handicapped, stay-at-home parents, part-time college students, and other high-promise segments. If you have to add staff to accomplish VAS, it's better to do it here than anywhere else in the organization.

The bottom line is this: Companies have spent millions (or billions) of dollars to build this incredible asset called the customer service and support organization. Then they take that asset and pump low-value content through it. I know that's hard to hear and you'd love to debate the point. But if value is about providing something useful that didn't exist before, helping customers restore access to the product or improve its performance doesn't create much value. It simply gives them back the value that they already paid for when they bought the product in the first place. Nothing more.

Imagine that you've just plunked down $70,000 for that zippy, high-performance Lexus discussed in Chapter 1. Unfortunately, performance degrades over time and has to be restored, parts need to be replaced, and cars need regular maintenance. Do these services add value beyond what the product could do on the first day you bought it? No. Is it running better than new? No. Are you able to do things that you couldn't do before? Not really. But what if the service also enabled you—for the first time—to access real-time traffic information on your navigation system, improve the car's gas mileage, and remotely uploaded your iPod songs into the car's audio system?

Now don't get us wrong. Maintaining system availability and performance *is* hugely important, and it's a responsibility we can never walk away from. But does it really *add value* to the customer? Is our current use of the customer support asset really the highest and best use from both a customer and shareholder perspective? We would strongly argue that the answer to both questions is a resounding no.

The trick is to provide services to customers that achieve the same growth stimulant as professional or consulting services but to do it in a higher margin way. This is the fundamental promise of value added services. And while user communities and Google may be handling the majority of customer questions today, they won't be able to deliver the next generation of services—at least not for some time. The vendor is the only one who can build the tools, aggregate the best practices, access the data, staff the experts, etc. It will be a big change. Even our partner ecosystem will be at a disadvantage initially. All but the very largest in the channel will lack the infrastructure and resources so critical to VAS success.

Fortunately there are some pioneering, real-world examples of technology companies who have crossed the line into VAS:[3]

- **Oracle Advanced Customer Services (ACS).** The single most economically important business to Oracle is its support and maintenance business and its 80%-plus gross margins. Maintaining high renewal rates is perhaps the most important single function of the company. But at the same time, Oracle also knew that driving increased customer value was becoming key. But how? Most companies have let this dilemma render them paralyzed, unable or unwilling to resolve the dichotomy between the growth effect of providing services and the lower margins that come along with that. But Oracle had an answer with its Advanced Customer Services offering. Here is the elevator pitch for ACS on its Web site: *Oracle Advanced Customer Services, a global business unit within Oracle Support, focuses exclusively on facilitating the continual operational improvement of your Oracle environment—throughout the life of your Oracle solution.* ACS is all about driving value from Oracle software by helping lower the customer's total cost of ownership (TCO) and accelerating technology adoption. And it is done from Oracle Support, not Oracle Consulting. Customers who buy ACS must already be customers of Oracle Premier Support—Oracle's highest level of support and maintenance. ACS gives Oracle the ability to provide high-value, high-touch additional services to its customers in a high gross margin way because ACS is built on top of, and is a part of, its customer support organization. Has this been a success for Oracle overall? According to Oracle's Web site, 94% of the S&P Global 100 has now purchased ACS. And most importantly, 100% of Oracle ACS customers renew their all-important annual Premier Support contract. ACS is an early stage but powerful example of VAS. Oracle has built a $500-million revenue stream for ACS alone, not to mention the impact it has had on growing its Premium Support. ACS gives Oracle an additional high-margin revenue stream, creates stickiness for its support and maintenance contracts, and enables consumption of existing licenses thus accelerating its largest customers' new license purchases. Customers love it.

- **HP SmartFriend** is a pioneering value added service offering in the consumer world for all things hardware and software. HP knows that the vast majority of support issues are not tied directly to its hardware, but to a combination of software, hardware, and services that make up the home computing environment. It also knows that support is the key to customer loyalty-tailored support, above and beyond traditional support. Here's how the HP Web site sells SmartFriend: *Have a security software question that needs answering, fast? Not sure how to install your new wired or wireless network, upgrade your operating system or keep your HP PC on speaking terms with all your other HP products? HP SmartFriend is the easy prepaid service designed to answer select hardware and software-related questions not covered by your standard HP warranty or addressed in your instruction manuals. HP SmartFriend is available by phone 24/7. When you need help fast, turn to a friend.* We have mentioned several times how important we see the Apple Genius Bar being. HP Smart-Friend is almost the offsite version. It promises fast, friendly, and reliable answers to your hardware and software questions for many programs—everything from Quicken to iTunes to WinDVD and even personal assistance with your digital photography, including tips and support for creating spectacular photos. Consumers receive one-on-one assistance from HP technicians purchased in single-use and six-month calling card options. With Smart-Friend, customers get the personal attention and tailored service they need without the hassle of leaving their home. Because HP SmartFriend is not proactive and not based on usage data gathered by the product, it does not qualify as a full-blown VAS example. But it IS targeting the right objective.

- **Philips Healthcare Utilization Services** is perhaps the best real-world example of VAS that we know of. It is available for its MRI and CT scanners used in hospitals around the world. Here is the selling pitch on the Philips Healthcare Web site: ***Benefits for you—*** *Use the information provided from MRI Utilization Services for guidance in defining and implementing improvement actions to: optimize workflow, increase patient throughput, decrease waiting lists, enhance quality control. MRI Utilization Services helps you identify what is going on in your department and how you are doing comparably ... Use the reports of your actual system utilization to identify opportunities for workflow optimization ... MRI Utilization Services gives you actionable direction on where and how to make improvements ... Enhance* productivity—Involve *your team to analyze information and implement specific solutions. Or invite Philips workflow experts to support you with best practice expertise.* This proactive service does everything we have talked about: It has metered the software in the product to gather usage data daily from every Philips scanner in the program. (It is an optional annuity service that hospitals can add to their maintenance contract.) It sends the data to an analysis tool that tracks all manner of usage patterns (time of day, procedure type, by operator, etc.) and compares it to other Philips scanners at the same hospital as well as other similar hospitals. The customer can see the results and performance comparisons the next day. Philips also studies the data to look for underperforming units. It then contacts that hospital to discuss the reasons and offer yet more services to help the customer increase the productivity of the unit through access to Philips' experts trained not in bits and bytes but in hospital workflow optimization. Philips reports that its Utilization Service customers are seeing two to three more patients per day (per scanner). Depending upon the reimbursement rate, that can be as much as 250,000 euro per year in revenue. This is a new offering—it is not yet six months old. Philips is currently tracking the product purchase history of Utilization Services customers to compare them to other hospitals who are not on the service. They have delivered the service in nine countries to date, and after the first-of-kind delivery in each country, requests in that market for utilization services has been tremendous, according to the company..

- **Salesforce.com Premier** is a great example of expanding a traditional support offering to take on end-user adoption as a core value proposition. Here is its Web site pitch: *Salesforce.com designed Premier Support to provide maximum value, a collaborative partnership, and personalized services. Our premier customer service representatives (CSRs) are the most knowledgeable Salesforce experts in the industry. They can tackle your toughest challenges, from troubleshooting for customers of all sizes to multithousand-user setups. Premier Support more than pays for itself in terms of higher user adoption, more CRM success, and increased business productivity. In fact, Premier Support customers have 20 to 30 percent higher log-in rates and benefit from 50 percent higher adoption of CRM features, on average. In addition to all the benefits of Basic Support, Premier Support includes:*

  o *A two-hour response time (business hours)*
  o *24/7 live phone support*
  o *An assigned CSR or team of CSRs*
  o *Health checks to determine if your company is using Salesforce to its full advantage*

- **Lawson Software Quickstep.** This is another early attempt at a VAS offering. Lawson sought to accelerate the delivery time frame for its ERP solution at each client while at the same time increasing the value derived from that solution through better end-user adoption. It accomplished this through a number of new approaches:

  o It verticalized its go-to-market approach, providing its solution in preconfigured packages so as to minimize the amount of configuration required.

- o It identified a best practices in accelerated deployments and standardized these into what is known as Lawson Quickstep implementation methodology.
- o It then sought solutions that would allow it to provide both greater automation and also deliver greater user adoption post go-live.

For the last point, Lawson partnered with a company called Datango to deliver the Lawson Learning Accelerator (LA). LA is a project accelerator that reduces redundancy of effort, that is, it is used during early implementation phases for documentation development (test scripts, etc.), reducing the time required to generate those materials. As a by-product, baseline educational materials are also automatically produced. Then they enhance these materials for end-user training with sound instructional design and complement them with standard best practices content that Lawson has developed. They leverage experiential learning/simulations in order to accelerate knowledge acquisition by the client's employees. Finally, Lawson delivers a comprehensive implementation that includes high-quality end-user training materials so that the client is implemented and up to speed faster, with its end users learning how to best optimize the system. The result is increased client ROI and reduced TCO, validated in a recent research study conducted by the Aberdeen Group that reaffirmed that Lawson's three-step approach translates to a faster return on the Lawson investment and more productive usage of the system—all of which (according to the company) contributes to Lawson having the lowest TCO among its competitors.

Every company will initially develop their own version of VAS. Not all of these will look the same. Not all of these will look like Oracle's ACS group. They all must be built and delivered in a manner reflecting their own economic and customer realities. Because of these economic limits there will likely continue to be a consumption gap around many products as there is a financial limit to how much VAS you can initially afford to provide. Remember, we are very early in tackling the efficiencies of this new service concept. It will improve as adoption of the model accelerates and as customers become more accustomed to paying for VAS services. The Technology Services

Industry Association will play a major role in identifying and reporting on VAS best practices and results across the industry. This will help member companies expedite their journey and minimize the costly trial and error of going it alone.

Not sure if VAS is the right offering for your company? You might try asking yourself these questions about what a successful VAS offering could mean.

- **Would your customers see it as delivering high value?** Giving customers more and better use of features is a value proposition they can definitely rally around, one that extends their current investment in technology to a higher level, and one that they are likely willing to pay you for. If your business customers' goal is cost savings, they get more of it. If their goal is competitive differentiation, they get more of that, too. Maybe your consumer customers' goal is to have photo albums on their coffee table for every vacation they take. You can help make that come true. These are service benefits that offer a clear and substantial return on investment to all manner of technology customer.

- **Are you willing to take on complex services and wrestle them to the ground?** Most customers will have a unique set of conditions that are slowing EUA. By applying the manufacturing principles already familiar to the support organization from their work in traditional availability support, aided by the coming breakthroughs in the emerging field of service science, you can begin the arduous process of breaking down some of the solutions to customers' complex EUA challenges into component information parts. With each new project, the library of service parts grows. They can then be recalled and strung together for use on another customer. Making this apply in complex, customized professional service engagements will indeed be a tough challenge but one ripe with opportunity, as we will discuss later in the book.

- **Can you deliver much of it out of your remote support centers?** Done right, a great deal of VAS value can be realized without anyone going on-site. That means the information needed by the customer can be identified, sent, discussed, and followed up on, all from the most efficient available location. So whether it's your call center in Austin or Ireland, Boston or Bombay, you can conduct the transaction from anywhere. In those cases where someone actually does need to go on-site, he or she could be dispatched from the field service organization or subcontracted to professional services staff or channel partners.

- **Could customers "self-serve" many of their VAS needs from your existing Web support front end?** Many of your company's existing Web service tools can be used to offer VAS content immediately. There is no need to buy much new. Customer service has state-of-the-art internal and external tools to disseminate specific information to a specific customer quickly and cheaply. Have you bought anything on a Web site recently? Did a little person pop up and talk to you? Maybe even walk around the site pointing things out? Imagine having that guided selling technology popping up to walk your customers through a VAS process.

- **Could you extend VAS services through your partner network?** Because of its robust process for capturing knowledge and making information available, your customer service organization can not only allow existing partners to access VAS information for their customers, it may also be able to give these partners a new revenue stream by offering EUA services. This is a major new opportunity for you to build positive but dependent relationships with your partners.

- **Could this become accretive to your professional service margins?** While many enterprise professional service projects are far too complicated to be delivered solely through a VAS model today, there are many subtasks that could. By developing a VAS model that includes information about the installation, integration, and deployment phases, the professional service teams could access valuable content for an on-site project faster and cheaper than ever. They may even be able to subcontract some professional services work to employees in remote solution/call centers. Both these activities have the potential to improve the historically weak project margins realized by many technology professional service organizations.

- **Could it be profitable?** Companies are likely to take divergent paths—at least initially—to get to revenue and profits through VAS. Some will bundle these services into their existing customer support and maintenance contracts to uphold their value and prevent discounting. Others will charge separately for VAS services as an additional service offering in their portfolio. (More about revenue models later.) But regardless of the method, we believe that VAS has the potential to be profitable as its own business in many tech markets because the ROI to customers is so substantial and easy to prove.

Did you respond to the questions more often with yes than no? Most companies do. It is a great vision but it is a considerable journey from here to there. That should not be an excuse for waiting. EVERY tech product has a consumption gap, and EVERY tech company needs to be developing some VAS offerings to shrink it. Even if you can't close every gap for every customer, we believe the net effect of what you do accomplish will be valuable to customers and accretive to the company's bottom line. What matters most is that you start your organization on the journey.

5 | Implications for the Organization

Wᴴᴵᴸᴇ VAS ᴏꜰꜰᴇʀꜱ ᴀ ɢᴀᴍᴇ-ᴄʜᴀɴɢɪɴɢ ᴏᴘᴘᴏʀᴛᴜɴɪᴛʏ ᴛᴏ ɪᴛꜱ ᴇᴀʀʟʏ adopters, implementing it will not be easy. There are real costs, real risks, and no shortage of land mines along the way. VAS is not yet a precise formula, and for the tech industry as a whole, truly achieving its potential will take time. Once the industry gets squarely into the VAS saddle, we expect that new tools, services, and processes will develop that will lower costs and improve effectiveness. But no matter how it plays out, the bottom line is that the journey is unavoidable if the tech industry is going to continue to grow.

So let's examine some of the major investments and organizational changes that are likely to be required, whether you're doing a small-scale test of VAS or rolling it out across your entire worldwide operation.

BARRIERS TO ADOPTION AND VAS-ENABLING THE PRODUCT

Like all good engineers or scientists, we want to begin our journey by understanding as much as we can about the issue at hand. In the case of VAS, the specific issue is our average customer's inability to take advantage of certain key feature sets in our products. Is this lack of adoption driven by a weakness in the feature? Not usually—for the most part, features inside products from major tech companies work pretty well. Could their designs be improved? Sure. But the main reason the consumption gap exists is that users are faced with too many complexities, with too many learning curves, and they have too little time or skill to address them.

As previously mentioned, the product managers we have talked to had definite notions about what their product's main adoption barriers were. So in the beginning, especially if the product is new, you'll have to go with your product managers' gut instincts. But since different customers will have different adoption barriers, the product needs to begin to talk to us.

More and more products are being engineered to track who is using them and how they are being used. This is only the beginning of the future of product metering. "Software metering" is discussed in Wikipedia as follows: *One of the main functions of most software metering programs is to keep track of the software usage statistics in an organization. This assists the IT departments in keeping track of licensed software, which is often from multiple software vendors. Desktop- or Network-based software metering packages can provide an "inventory" of software, give details of all the software installed in the network with the total number of copies with the usage details of each software, and even track metrics of software use such as how often it's used by a particular department, the peak times it's being utilized, and what add-ons are being utilized with it.*[1] Moving this concept along to the user and feature level is already a reality and should be a standard practice in the coming years. Most technology product categories already possess the few attributes that make metering possible. They almost always:

- Have a software component to the product.
- Know who the user is (either assumed from registration on single-user systems and Web sites, or by access grant on multi-user systems).
- Have the ability to "phone home" via the Internet.

Take any product with those three attributes, and you have the potential to make VAS metering not only a reality, but affordable. If your product has those attributes, there's a good chance that it already incorporates many of the capabilities that will allow you to record and aggregate information about how the product is or is not being used, and by whom. Autodesk has made that investment, as has Microsoft and many others. Ultimately you will combine that information with some psychographic, demographic, or firmographic data from your CRM system to give you better insight into the different adoption barriers facing different customer or user segments.

If you aren't already metering on your own, the good news is that there are a growing number of third-party software products that can be imbedded into your product, which give you the capability to track and report on EUA. But even with that, it's going to take commitment. Besides the financial cost, adding additional metering code may also slow your product's development time or its performance. Thus you will have to win over some mindshare in the development organization. But it is essential for VAS to become practical and affordable. You don't want to hire teams of people to manually assess how customers are doing everyday when the product is fully capable of measuring its own usage and reporting it back to the company. It's not that VAS cannot be delivered without this information; it just makes the whole process more efficient and effective.

One of the earliest attempts to meter a product for service purposes was a BEA Software offering called Guardian (now owned by Oracle). What Guardian did so well was to capture the status of the BEA product and the surrounding technology whenever a problem occurred and send that information home via the Internet (in much the same way as Microsoft does whenever one of its programs crashes). When BEA began to notice a pattern of conditions in or around its product that consistently created certain problems, it identified them as a "signature." BEA attached to the signature a set of instructions (do this, download that, etc.) that the customer, BEA support, or the BEA product itself needed to do to resolve the problem. Once distributed to a customer, if their system detected similar conditions present on the customer's systems, it would alert the customer and BEA, and begin executing its instructions for resolving the problem. Signatures could be created by BEA or the customer. BEA would aggregate and QC them and then distribute them to all customers who purchased Guardian. Guardian's net effect was a quantum reduction in the number of high-priority support cases from those customers who installed it. And it was all done by the product itself, along with a small number of support engineers who oversaw the quality of the signatures. While we are proposing collecting product data for a different purpose (value realization vs. system availability), it was a forerunner of what is to come as it incorporated a number of the attributes that we believe are essential to creating VAS. It sat on the customer's server and monitored how the product was used and in what context. It then sent the data home to BEA where it could be aggregated and analyzed, and where service "parts" could be attached to the prescriptive part of each signature. Many of the service parts were automated so that neither the customer nor BEA had to apply labor to the problem. Signatures could be created by BEA or by the customers themselves. At some point the majority of Guardian's ongoing signature development costs were borne by the customers.

Once VAS metering is imbedded in a product and you begin collecting real data on EUA, you will need a smart way to turn those mounds of information into something of value to you and your customers.

VALUE MODELS (VMs)

To do that, you'll build something we call *value models (VMs)*. The concept of VMs is simple: Define a related set of features that, when used properly, deliver uniquely high value to a particular segment of customers. Then for that set of features, identify, package, and deliver the support assistance, training, or business process recommendations a customer needs to make its end users successful at realizing those features' value.

At least conceptually, IBM Redbooks are a low-tech, nonautomated example of what an end-user value model looks like. (If you're in enterprise computing, you're probably familiar with these.) Here is how IBM describes the 3,000+ Redbooks that are available on its Web site:[2]

> *IBM Redbooks are developed and published by the IBM International Technical Support Organization, ITSO. The ITSO develops and delivers skills, technical know-how, and materials to IBM technical professionals, Business Partners, clients, and the marketplace in general. IBM Redbooks are the ITSO's core product. They typically provide positioning and value guidance, installation and implementation experiences, typical solution scenarios, and step-by-step "how-to" guidelines. They often include sample code and other support materials that are also available as downloads from this site.*

Your company probably already has some documents that resemble Redbooks. But why does IBM have 3,000 and counting? Not only do large companies have a lot of products, but they also face the challenge that one feature set might need multiple VMs, depending on whether the same features are used differently from industry to industry or from usage type to usage type. A server product may have two different ideal deployments of disk partitioning, depending on whether its primary goal is to deliver speed for transaction processing or security for large file data storage. A particular feature set inside a software application might require more than one VM if it is used differently in manufacturing companies than in services companies. A phone system might best use its ringtone capability differently in a hospital than it does at a retailer. In other words, you could have VMs for the same product that differ by SIC code, customer application, or any number of other considerations. But in any case, like BEA's signatures and IBM's Redbooks, what is critical about our value models is that they act as the key instruction set for VAS.

Once you've targeted a set of features and customers who warrant a VM, you must then put together what, in effect, amounts to a bill of materials (BOM). But unlike a classic manufacturing BOM, ours is not made up of physical parts. Our VM BOM is composed of *service* components. These service "parts" could come from anywhere in the customer service, field service, professional service, or training departments. There may even be some you have to build on the fly to fill key gaps in your inventory of service parts. Before we are through, we will need to translate these service parts into typical training elements and tools such as knowledge articles, tutorials, templates, online rich media, software scripts, reports, and anything else that could help a small group of users get up to speed on a specific set of high-value features or amend a business process. While ideally all of these could be delivered remotely and electronically, some of these service parts may require human assistance from a customer service center, or even on-site assistance that could come from your field services team, professional services team, training team, or a third-party channel partner. But if it is required to deliver the value model, then it goes on the BOM and the VM BOM is priced accordingly.

Because there are so many possible VMs on a complex product that you're selling in multiple markets, your company will have to be smart about which ones you take the time to focus on first. As you build your list of VMs, the product's potential to deliver actual value to the customer base increases. Remember that the value models could consist of elements of any or all of your existing service offerings depending on what is needed for the customers to succeed on a particular feature set.

Think of VM BOMs as short-duration service "mini-projects" that will be delivered to a particular customer from across all the key service organizations within your company and your partner ecosystem. In the enterprise space, they will usually need to be managed like projects because they will involve more end users and a more complex set of tasks, but hopefully you will design the VM BOMs with the tools to allow remote project management from one of your customer support centers, not from an on-site project manager.

Although enterprise companies face additional VAS complexities, they will likely have the easiest time building the initial business case for VAS. They will be most likely to have the revenues to support a medium—or even high—initial labor component. This will allow them to enter into the VAS market faster and with more immediate effect than their peers selling high-volume, low-cost consumer technology products. For those companies, there will be far more pressure to design, develop, and implement technologies rather than people to deliver VAS. As we already pointed out, it will take some time before VAS enters the Age of Tools. The good news is that we expect it to move there faster than any other service in history. All the tools are there; they just need to be redirected to this purpose. And finally, let's not lose sight of the end goal of these VAS services: to ultimately sunset themselves back into a future version of the product.

So you should prepare for a period of somewhat higher costs as we begin the transition to VAS. But importantly, this does not necessarily mean radically lower margins. Later we'll discuss how you can use this to create a true source of competitive differentiation in a time when services are becoming important criteria for vendor assessment by both consumers and businesses. We'll also review how VAS can improve your product margins and reduce your cost of sales, especially to existing customers. Finally—and most importantly—we'll discuss how VAS can accelerate customer repurchases. That alone could pay for your company's entire value added service journey. And, oh yes, there will also be new service revenue to go along with that.

THE CONVERGED SERVICES ORGANIZATION

For the first time ever, technology and infrastructure are emerging to tackle the consumption gap and to do it profitably on a large scale. If you look around, you'll see pieces of VAS emerging everywhere. But few companies have put all the pieces together to create a complete VAS program. That's because to do so requires some major organizational changes.

The most dramatic of organizational changes is something we at TSIA call "services convergence." In our context, the word *convergence* means bringing together the entire professional service, customer support, field service and education divisions into a single-delivery organization. We are not talking about simply having all of these siloed organizations report to the same senior executive, but of truly integrating these divisions (plus other service functions such as managed services or in-store services) and leveraging each one's unique expertise to create a more powerful, more integrated, and more profitable converged services business.

While convergence will be a more complex scenario for large enterprise tech companies where big professional services organizations exist, it is also a part of the future for consumer and SMB tech companies. In those cases, the convergence may be between remote technical support centers and in-store support. For most tech companies that have their own retail locations, those two organizations are in completely separate divisions of the company. Many companies also have customer training divisions and/or channel partners who perform that role and need to be brought together. So while convergence will not look the same in every company, the reality is that we will all have to perform similar organizational gymnastics to deliver VAS.

Some CXOs may face a bit of denial about the current state of integration between their services LOBs (lines of business). They may appear integrated on the company Web site, in the sales pitch, or even at some level of the organization chart. But if you look inside most tech companies today, the services are not really integrated. Each of these LOBs usually has a separate P&L statement, business strategy, and management at the global, regional, and country levels. They frequently use their own separate IT tools and make pricing and packaging decisions independently. Too often they are sold through separate sales processes and appear to the customer as autonomous (and frequently poorly communicating) organizations. As we will discuss in Chapter 6, most of this is done to give the company the maximum management and P&L accountability for each LOB executive.

At TSIA we work daily with the separate service silos of our hundreds of member companies in the IT, medical, automation, and many other technology industries. Member organizations frequently find out information about another service silo's internal strategy or practices within their own company *from us!*—that's how siloed they are. Given the traditional management structure of most companies today, it's easy to see how this situation might develop. But here are four good reasons why you absolutely must break down these silos:

- The individual silos don't benefit from the core capabilities and expertise of the others.
- Customers receive fragmented services from the company, rarely receiving an integrated services plan.
- VAS will require the resources and talents of all the silos.
- There are numerous cost efficiencies to be gained.

Let's take some time and examine the first point. In our work with companies, we see some clear and compelling areas of expertise being formed in each of the four main services silos. Here is how we would describe them:

Professional Services: The application and engagement experts. No one in your organization spends more time helping customers actually move along the adoption life cycle with your products. No one knows better what customers actually want and try to do with your products. They want to deploy it in a certain way, integrate it with certain products, achieve certain business results and need help adapting certain business processes—all in an effort to achieve a constantly evolving set of business objectives. Because these are often unique and difficult, the professional services organization must be staffed with smart, creative engineers and experts. They craft clever solutions that give the customer the application of the product that they want. They know what the gap is between what the customers want and what the product can do. The professional services people are also the most expert in managing the vagaries of the complex project management process. As our engagement experts, they know where the profitability land mines are in a typical project and what tools and techniques should be used to manage around them. They also understand the details of customer expectation management, the change order process, and dozens of other engagement practices that can spell the difference between profitable and unprofitable project work. And finally, of all the service silos, they usually have the highest level of industry and customer domain knowledge.

Tech Support: The service manufacturing experts. These people know how to take a service, break it down into components, turn them into digital "parts," make those parts efficient and cost-effective to reuse, get them into a global production system, and then combine them in a live production environment to meet a particular customer's needs. They make customer service experiences that are effective, low-cost, easy to scale, and consistent around the world. They can even take a complex situation like a multi-vendor customer service problem and turn it into a series of knowledge articles that can be consumed instantly by tech reps—or directly by customers.

Field Service: The field efficiency experts. No other customer-facing organization in the company can get to the customer's site faster, with better tracking, with more quality control, or at lower cost. They know the exact location of every engineer at any time, and they have technology to optimize workload and minimize downtime. They work seamlessly with the tech support organization not only for dispatch but also for providing information about the case while it is open and reporting the details after it is resolved. They have also mastered global parts logistics and reverse logistics (bringing parts and products back from the customer) that get the right parts to the right field engineer at the right time at the right location. They can do that efficiently in Chicago or Cairo, Paris or Penang. Together, tech support and field services are the one-two punch of service efficiency.

Training and Education: The learning experts.   Need to get 10,000 customers of all abilities and experience levels to learn a new concept or redesign a common business process? Who are you going to turn to? Your training and education team stays on top of leading-edge education principles, learning behaviors, content creation, and delivery techniques. If you need to bet on someone in the company who can optimize the consumability of information or knowledge, bet on these people.

So here we sit as tech companies with these four valuable competencies completely segregated in the different service silos. And worse, these are all too often separated from the critical development and sales silos. (Again, your silos may be slightly different.)

In any case we need new thinking about the services organizational model, because to get to VAS, we need a highly scalable, global operation that can cost-effectively deliver mass-customized projects to the customer. We need an operation that combines the best attributes of each of the current silos.

How do you begin to attack such a daunting organizational challenge? Ask yourself this very basic question: What does it take to deliver a service—any service? One unique aspect that all services have—one that separates them from many other business processes—is that they are co-created in real time with the customer. Unlike building a product, where the development and manufacturing environments are relatively static and you control every aspect of production, creating each service is unique. Each one is designed and delivered based on unique information from and insight by the customer. The answers are generally co-created through a vigorous dialogue between the customer (or the customer's unique technology environment) and your employees and their tools. Thus we all share the same service goal of "co-creating a solution with a single customer and rolling it out to other customers successfully at the lowest possible cost." By definition, then, all services should go through the following four steps:

Step 1: **Solve.** Use employees to co-create the solution (or the value) with the first customer.

Step 2: **Productize.** Identify the existing service "parts" that were used in the solution. Identify new "parts" that need to be created.

Step 3: **Package.** Make these reusable "service parts" as easy as possible for other customers or employees to consume.

Step 4: **Scale.** Get the solution out to every customer and employee through the most cost-effective means available.

You may have different words for the steps internally, but the same principles apply. In the emerging world of services research and services systems design, these steps can be seen as both "front stage" or "back stage" components. As UC Berkley professors Bob Glushko and Lindsay Tabas explain:

> *Service management and design has thus far primarily focused on the interactions between employees and customers. This perspective holds that the quality of the "service experience" is determined by the customer during this final "service encounter" that takes place in the "front stage." This emphasis discounts the contribution of the activities in the "back stage" of the service value chain where materials or information needed by the front stage are processed. However, the vast increase in Web-driven consumer self-service applications and other automated services requires new thinking about service design and service quality. It is essential to consider the entire network of services that comprise the back and front stages as complementary parts of a "service system." We need new concepts and methods in service design that recognize how back-stage information and processes can improve the front-stage experience.[3]*

Our Step 1 is the classic front-stage activity. But because we desire to roll it out to all customers successfully tomorrow at the lowest possible cost, we also engage in Steps 2–4 as our back-stage activities. This is a premise that is simple to envision.

Here is the problem with our current approach to services. Each of the four major service silos in our companies engage in both front- and back-stage service activities *independently of the others.*

Current Service Delivery

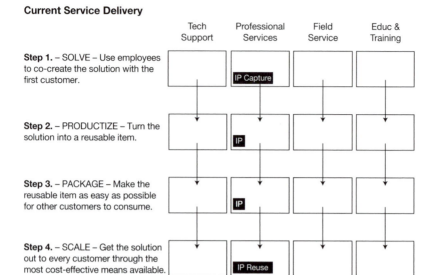

FIGURE 5.1 Current Service Delivery.

And quite frankly, they are not all equally adept at each step.

Current Strength & Weakness

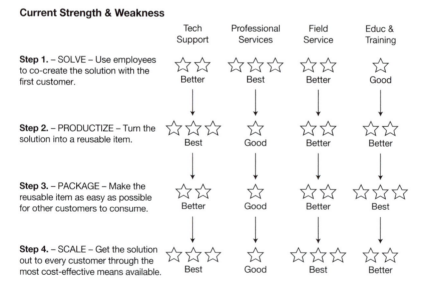

FIGURE 5.2 Current Strength and Weakness.

In Step 1, from a VAS effectiveness perspective, the three other service organizations get the edge over the education and training organizations simply because they are co-creating new solutions with customers each and every day. They know what is happening with individual customers on a real-time basis—what is working, what is not, and what remains to be done. But even here these three organizations are not all created equal, especially not in the world of VAS. This is because the nature and value of the solutions they provide are not of equal value to the customer.

Remember our customer adoption life cycle chart? In that model we argued that the further to the right that your services move, the more value they create for the customer. While maintaining or restoring availability indeed has value, it does not add anything beyond what the product could originally do. Thus the tech support and field service organizations are edged out by the professional services organization simply because the value advancement opportunities (moving customers to the right on the adoption cycle) are greater.

In Step 2, *productize*, the tech support organization has the advantage. These employees know how to take a complex solution and break it down into steps. They create more new formal knowledge for their knowledge base every day than any other organization. They know how to make that knowledge findable by any employee or customer. They manage it using the most sophisticated tools in the world. In many companies the tech support organization is already managing or jointly managing the knowledge from the field service organization.

In Step 3, *packaging* (which we defined as taking the reusable service solution and making it as easy as possible for other customers or other employees to consume), one organization again has a clear advantage over the others. The education and training organization is staffed with content development experts who are trained to do this for a living. Many are not engineers, and that may be a particular advantage in the VAS world where the service solutions will be much more end user-oriented. While all four organizations do some amount of knowledge triage in order to make content ready for consumption by other employees or directly by the customer, the true knowledge creation experts are in education and training.

So there are different "experts" at different points in the service process. But even before we talk about the implication on VAS, let's talk about how you could begin your convergence journey just for the sake of cost savings and increased efficiency. How? By centralizing the middle two steps. Take the key people and technology needed to productize and package solution content and bring them together in a converged function. This will likely be managed by people who currently reside in the tech support and education silos because of their deep experience levels at these disciplines.

Converged Service Delivery

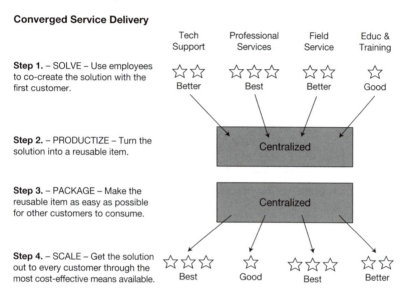

FIGURE 5.3 Converged Service Delivery.

First of all, by converging these two functions you may be able to eliminate some redundancies in people or technology. More importantly, you will extend the knowledge management (KM) culture that currently resides in the tech support silo to professional services. Inside this culture are proven business processes and tools that the industry has now spent a decade improving—processes that can and should be extended across all the silos. You will need aligned strategies and actions across the silos to do this—all the way down to how we compensate our people. This one convergence step could have significant ramifications on improving the economics of the professional services business. Not only is that one of our current challenges, but it also takes pressure off of the decline in overall service margins as the mix of services shifts away from annuity services and toward project-based services.

Of course, it is also an essential step as we move toward VAS. In a true VAS world we will need a plethora of service "parts" to make up our VM BOM. These parts have to be created from every service silo and for every stage of the customer adoption life cycle from purchase through value realization. Most of these parts will need to be built to be consumed in a wide variety of modalities. A single service "part" may need to be authored into a Web-consumable article, a tutorial, a classroom slide, a documentation addition, and a self-help pop-up. Software will need developed and tested. Depending on the price of your product, the size or location of the customer, or a number of other factors, you might need to create both low-cost, remote parts as well as high-touch, on-site versions of the same parts. In short, you need a knowledge management factory, which will become the center of your new "converged" services business. You need the best and brightest in the subdisciplines of KM to be working together to provide a globally accessible, timely, multi-modality, cost-efficient knowledge base that spans every phase of your customer adoption life cycle model.

Given our proposal to repurpose your tech support infrastructure to become the manufacturing backbone of the new VAS delivery organization, you end up with a VAS services model that looks like Figure 5.4.

VAS Delivery

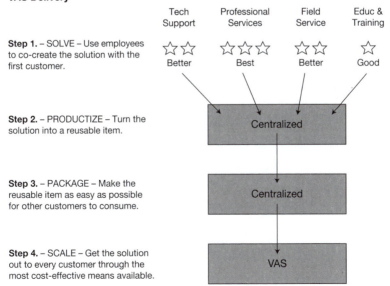

FIGURE 5.4 VAS Delivery.

As a final benefit of the model, we believe that your professional services organization will begin to pass a part of every project onto the VAS platform. Once eligible components of the solutions from the professional services field organization become part of the universal knowledge base, a future service engagement can be divided up. Your professional service team will have to create new service solutions—usually on-site—for the unique aspects of that project. They will remain costly and complex. But perhaps a part of the total engagement can be delivered remotely through the VAS organization. This should make the overall engagement more profitable than current models.

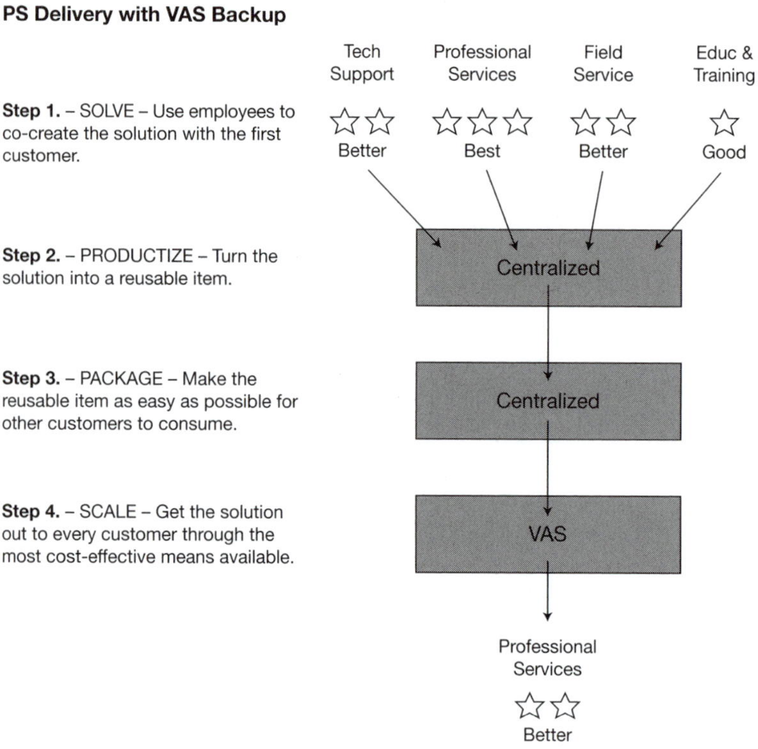

PS Delivery with VAS Backup

FIGURE 5.5 PS Delivery with VAS Backup.

It's easy to brush this idea off by saying that professional service tasks are simply too complex, custom, or technical to be broken down and relegated to things like a knowledge base. For some tasks that will be true, but not all. TSIA member companies are already reporting that about 20% of professional service projects today involve allocating hours to solution center staff (a centralized professional service facility where particular work can be done by centralized experts more efficiently than in the field). And encouragingly, those projects are averaging a gross margin in excess of 40%, compared to the average of projects that do not utilize solutions centers, which average just over 30%.[4] That is over $100,000 more profit on a $1 million professional services project!

In fact CA (Computer Associates) is doing this today. It has teams in the customer service organization that work with CA Services (its professional services group) and its largest channel partners to make end-customer professional service projects operate faster, more smoothly, and at a higher margin.

Similar "It's too complex..." arguments were made about customer service problems in the 1990s. But today's customer service organization has proven that it has the tenacity and tools to wrestle many complicated problems to the ground. Even if initially just 10% of professional service tasks can be offloaded to a remote location or can be delivered by lower cost labor because the information resides in a knowledge base, it could make a real dent in the project profitability dilemma that it too often faces. Give it a try. Take a couple of reoccurring professional service tasks and see if the customer service team can develop some KM content that aides the field professional service teams or allows some work to be done remotely.

A final reason to select these two steps for convergence is the beneficial relationship the new knowledge factory could have with the product development organizations.

A few pages ago we stated that the traditional service goal has been *co-creating a solution with a single customer and rolling it out to all customers successfully at the lowest possible cost.* But more recently a number of progressive companies have applied the Third Law and updated their goal for services by further refining the description of *how* the solution should be rolled out to the customers.

At TSIA we think the optimum description of tech services will soon be this: *Service is a thin veneer of smart people who get learning from customer advances and speed them through one-to-many systems and ultimately into product improvements.*

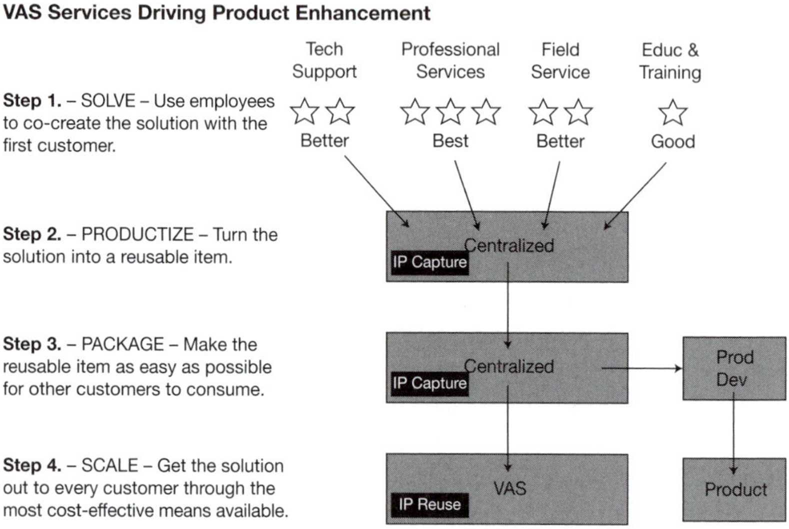

VAS Services Driving Product Enhancement

FIGURE 5.6 VAS Services Driving Product Enhancement.

If any of the four attributes:

1. A thin veneer of smart people, not a thick one

2. Learning from customer advances; that is, co-creating value with customers

3. Through one-to-many systems (KM, Web sites, tutorials, VM BOMs, etc.)

4. Ultimately into the product

… are missing from your execution, then you are spending too much, and doing a disservice to customers. This is particularly true of traditional service offerings (maintenance, implementation, etc.) that have *fully* entered the Age of Product phase where—by design—they should be getting "sunsetted" as the product now does what the services previously did.

So the future state of tech services is simple enough to envision:

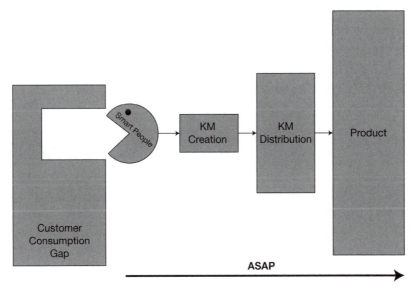

FIGURE 5.7 The Future State of Tech Services.

Take Oracle's Advanced Customer Support offering. One reason why Oracle ACS is so groundbreaking is that, in its model, ACS engineers not only specify process changes for the customer and for Oracle Support aimed at reducing the customer's total cost of ownership, but they also have close linkages with key product development resources to drive their learning into specific product changes. This is good for both the customer and Oracle.

In conclusion, the overall concept of services convergence is simple to understand but hard to execute. Nonetheless, tech companies must begin the journey. They must find ways to extend the existing strengths and assets of each of the silos to the entire services organization. These efforts will benefit both customers and shareholders. Services convergence has the potential to revolutionize the customer experience, reduce the consumption gap, and improve service margins in customer support, professional services, outsourcing, and managed services.

STEPS TO DEVELOPING A VAS PLAN

But where do we start our journey toward VAS and convergence? Let's begin by drawing a picture. What might your early stage, converged services organization look like? Conceptually, the converged organization should share a common infrastructure that looks something like Figure 5.8.

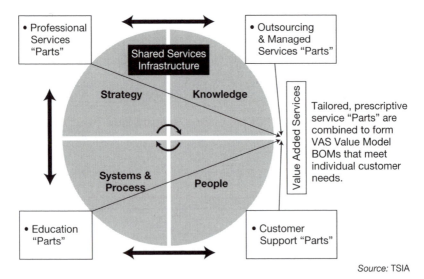

Source: TSIA

FIGURE 5.8 Service Organizations of the Future Will Look More Like This.

Most importantly, there will be a single unified strategy that spans the entire services portfolio, not one per silo as it often is today. That is our first deliverable. The strategy should:

1. Identify, rationalize, and retain the most important and effective offerings and strategies from each of the existing silos. (Because those offerings are not going away!)

2. Lay out a list of VAS priorities—those product adoption barriers for specific customer segments where a successful VAS offering could best advance the company's market position and financial performance.

3. Define the first set of VAS offerings—a preliminary set of value models that you wish to offer based on the product adoption barriers that you defined in Step 2. Combined with your offerings from

Step 1, you now have the first pass at your rationalized, converged service offerings—both traditional and VAS.

4. Armed with this list of offerings, you can begin the fourth stage of inventorying and assessing the assets (capabilities) you currently own across the service silos. There are two goals to this step. One is to look for redundancies that can be eliminated. The second is to inventory all the service "parts" that you have or could build. Remember that a goal of your strategy is to create nonoverlapping service "parts" from each discipline that can be combined to deliver a custom services solution to the customer, VAS or otherwise.

5. Armed with your lists of offerings and service parts, you can begin to build your first generation of VM BOMs. This is where it all comes together. This exercise will tell you a lot about where you stand:

   a. Are there existing service offerings whose margins could be improved by a more efficient use of existing parts from other silos?

   b. What VAS VMs could you offer today out of your current inventory of service parts?

   c. What would their costs be, and what would you need to sell them for?

   d. What are the new service capabilities and parts you need to create or add to make more of your desired VAS offerings possible for your target products?

   e. What will those additional costs be, and what would you need to sell them for?

   f. What role do you want your channel to play? How much of the channel is able to deliver certain service parts today, and how will you work with them?

Taken together—pending the obvious steps of customer demand validation of offerings—you have the framework for your new services strategy. The strategy document will outline a potential set of offerings for both VAS and traditional services, as well as a more unified strategy for systems, processes, and knowledge management across the silos and the partner ecosystem. You will really begin to "see" how things need to come together.

As a by-product of this exercise, you will likely find that you have many duplicate investments across the service silos. As an example, we have already discussed the many advantages of centralizing the KM functions. But going further, we believe that most all service offerings should be built on top of a common systems and tools platform. How much waste occurs today because of redundancy in IT spend across these silos? It is hard to imagine. And just as it is with KM, not all the silos are equally adept at their commitment to and benefit from IT. Tech support IT has a strong mastery of CRM, KM, and Web collaboration tools. Field service IT knows the most about mobile devices, parts inventory and logistics applications, and schedule optimization tools. Education knows learning management, Web-based training/e-learning, and lab-based simulation tools. Professional services are using some form of PSA tools to manage their engagements, resource allocation, and project billing.

You may want to form a cross-silo task force to conduct this assessment. Ultimately that task force might evolve into a centralized IT/ CIO function to both reduce costs and better leverage these unique skills. And then, what about HR? What about services marketing? There is a lot of fertile soil in these types of hard questions.

Now here is the caveat emptor: In our work with our primary consulting partner PRTM, we have found that it takes senior-level buy-in before you can even begin this strategy effort. Since you will be treading on a lot of different internal kingdoms, you better have top-level support and buy-in—not just among the service silo executives, but also among the heads of sales and product development, maybe even the CEO. It also helps to have an independent third party to move the project along and arbitrate disputes. You may break a lot of glass, but remember, you are doing it in the best interest of customers and shareholders.

Following the development of your converged service strategy document, you may elect to take the first steps of organizational convergence like those defined above. That's great.

But eventually, in perhaps the most challenging phase of implementation, you will need to open the door between people in the silos to better let the expertise flow. This is a cultural "whopper" of a task. At some point you must take on the arduous task of breaking down the separate service silo walls and slowly beginning a journey to allow human resources to be drawn from almost anywhere to best match the current shape of service demand across the converged organization. As an example, the professional services team always seems to have a group of non-billable talent (people who are not currently assigned to a fee-based project or are unable to be billable due to delays on the project). Collectively these people are often said to be "on the bench," meaning that their particular skills are not needed for today's current portfolio of engagements. Minimizing time on the bench is always a key to professional services profits. But tomorrow morning a deal might get sold that renders professional services short of that same skill set. Imagine trying to handle staffing in this ever-changing mix of customer projects! They need 23 people with skill X but they have 35 on the bench, so they have overcapacity. Then next week they need 42 people and they still have 35. Now they have undercapacity. Often they will go to external sources to procure the extra talent (at extra cost). But wait! Across the silos, both technical support and field services have technical resources that are either off peak demand or, in the case of field services, are looking to be revenue-generating.

We believe that there are some limited synergies here through convergence. By managing the services labor pool in a more holistic way, we can improve our labor utilization. Not all employees or all tasks can be freely exchanged between functions, but clearly some can. A Level 2 product specialist in tech support at the Dublin call center may be able to add a lot of value to a professional services team working on a complex deployment of that same product in Dubai. A field service rep in Seattle may be able to bill eight hours a week for the next month on relatively rote tasks needed on a professional services project in town. Those professional services people who are "on the bench" may be useful to the education teams in either curriculum development or custom training opportunities. In a retail environment, maybe a "services concierge" could physically sit a customer down for a training session in the store but turn the actual training over to a tech support employee in a remote call center. The point is that we can utilize a universally managed labor pool to reduce some labor inefficiencies around the edges of all the service silos. This is certainly not possible in all cases or for all job positions. But it won't take much of an improvement to add up to significant savings. If you could just redeploy 2% of your total service labor hours, it would save the need to add more employees as you grow.

If you are not yet ready to go for a complete strategy, Figure 5.9 lists nine specific areas where TSIA believes that you could begin to create synergies by converging the service silos today through more targeted projects.

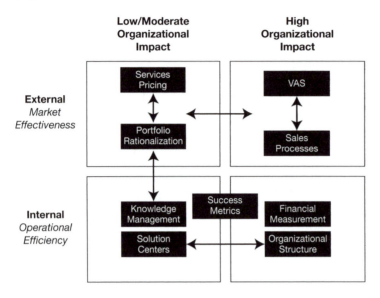

FIGURE 5.9 Service Convergence Initiatives.

Let's take a further look these areas and their relative degree of organizational implementation challenge.

Convergence Initiatives

Area	Objective	Target Benefits	Organizational Challenge
Services Portfolio Rationalization	*Reconcile service offerings across services organizations*	• Reduced spending on services development • Reduced customer confusion	Low/Moderate
Centralized KM and IT	*Create common IT and knowledge management infrastructure across service services organizations*	• Reduce cost of IT • Increase leverage of knowledge assets • Improve PS margins	Low/Moderate

Area	Objective	Target Benefits	Organizational Challenge
PS Solution Centers In CS	*Leverage centralized resource pools in CS and PS*	• Reduced labor costs • Increased scalability of skill sets • Decrease need for additional overhead	Low/Moderate
Organizational Convergence	*Combine existing services organizations*	• Reduced expenses • Increased coordination • VAS enablement	High
Consolidated Services P&L	*Break down current financial stove pipes between CS and PS offerings*	• Reduce internal financial reporting requirements • Accelerate ability to create mixed offerings • Force further refinement of VSOE to combined service offerings	High
Bundled Services Pricing	*Simplify and combine services pricing across CS, PS, MS offerings*	• Accelerate account adoption of services offerings	Low/Moderate
Total Customer Profitability Reporting	*Redefine success metrics beyond services revenues and margins into customer profitability*	• Understand impact of services activities on account and product success	Low/Moderate
Integrated Service Sales Processes	*Sales structure, processes, and skills to support selling of complete services portfolio*	• Reduced sales expenses • Increased account revenues • Increased account profitability and stability	High
Value Added Services Portfolio	*Development and deployment of VAS service offerings designed to accelerate product adoption*	• Increased account revenues • Increased product adoption • Reduced customer churn	High

At TSIA we believe that services convergence is not only a huge opportunity to save money by eliminating redundancies across the silos, but it is a critical step on the journey toward VAS. You will have a more efficient and more capable organization than the sum of the four disparate silos today. But you need a rock-solid foundation—especially in your current customer service operations. Since the customer service infrastructure will likely be the "manufacturing core" of your converged infrastructure, what you start with can't be weak. For companies that are missing some or all of the elements of world-class customer service, the first step of your journey is pretty obvious.

Fortunately the industry's best-practices processes and tools have been captured and are available to your support executives through things like the TSIA's *Organizational Development Program (ODP)*. Road maps such as these provide a tried-and-true bet to companies who want to fast-track customer service effectiveness either for VAS or just to improve basic customer satisfaction or support efficiency. There is no need to endure the costs and frustrations of figuring it out on your own.

At leading companies like Alcatel, Avaya, BMC Software, and CA, the "convergence" movement is underway. A simple "first step" example is the CA Customer Value Network. CA is creating a united service brand and experience that spans all the service silos. Here is its Web site pitch:[5] *Achieve success with your CA Solutions by taking full advantage of the programs and offerings available to you in the CA Customer Value Network (CVN). The CVN brings together CA Services, Education, Support, Communities and Partners to deliver programs, communications, events and offerings to you. Whether you are a new CA Customer or have extensive experience with CA Solutions, you can educate yourself on new product and technology updates, get in-depth CA Solution technical knowledge, share ideas and best practices with your peers and more.*

- *Shorten time to business results and reduce your risk with CA Services.*

- *Obtain the training you need to maximize your software investment with CA Education.*

- *Maximize the value of your CA Solution with technical expertise and self-service tools from CA Support.*

- *Gain access to CA planning and insights, influence product development and interact with your peers through CA Communities.*

- *Work with CA Partners who provide technical and service expertise that complements CA offerings.*
- *Increase your ROI in CA Solutions by engaging with CA in the programs and offerings available in the CA CVN.*

CA is not yet fully converged, nor is it truly delivering VAS, but it is positioning customers—and employees—for that day.

We are not trying to hide the difficulty of convergence and VAS. It will be gut-wrenching. But we are trying to provide you with the evidence that this is the future of technology services. These are indeed big decisions with multiple strategic, financial, and cultural ramifications. Convergence and VAS will mean something different to your company than to others. What you can converge to make VAS work, how fast, and to what effect are all idiosyncratic of your unique situation. And clearly there will be a different model for consumer companies than enterprise companies. But in any case, the direction that convergence sets you off on is right for customers and shareholders.

A SEA CHANGE FOR THE CUSTOMER SERVICE ORGANIZATION

As the company's tech support call centers and field service technicians begin to shift the nature of their work away from break-fix and toward VAS, a profound evolution in human resources must take place. That's true from the first-line rep to the senior vice president of support. For years, technical support organizations have prioritized technical knowledge in their staff. The more technical certifications, the more code they know, the better. Major offshore destinations for tech support call centers, like India, have made massive investments in universities that focus on technical education. The idea is to produce an ever larger pool of technical experts at low costs to attract major technology companies to relocate their call centers or to choose outsourcing providers in that country. While this has clearly been a successful strategy so far, an industry move toward VAS may make them shift their focus.

We would go so far as to say that offshore support destination countries should begin to rethink their workforce development strategies. They must provide not just graduates with deep technical knowledge, but graduates with application expertise as well. Because even more than technical skills, VAS delivery will reward knowledge of the customer's domain, application expertise, and, in business technology markets, their industry—basically the desired usage of the product. In enterprise tech companies, that means you might need banking experts talking to banks, security experts talking to IT managers, manufacturing experts talking on the floor to manufacturing managers, or nurses talking to doctors. In fact most enterprise tech companies have long seen the need for vertical focus. All enterprise application software companies have at least some vertical organizations. They (and other tech companies) might have vertical marketing, sales, or professional services organizations. What is the one customer-facing organization that has never been verticalized? Customer service and support. Now is their time. We need what some experts call "T-shaped people." IBM describes T-shaped people like this:

> *The future of business demands a new breed of knowledge worker: the T-shaped person who combines broad understanding of business processes (the top, horizontal part of the T) with deep practical execution in a specific functional area (the bottom, vertical part of the T). People who share the same understanding of the business process (top of the T) can team with colleagues with different I-shaped specialties (bottom of the T) to cover the waterfront of a business need without losing that common vocabulary and understanding of their shared business objective.*

> *IBM: Future of Business — Steve Mills — June 2007[6]*

In the specific case of VAS, the ideal customer service representative is one who has a broad understanding of technology, has good general business and personal skills (both a part of the horizontal part of the T), but is deep in an area of the customer's business (the bottom, vertical part of the T). Another change that will accompany this shift is new forms of certification for service talent. Right now we depend almost entirely on technical certifications. We use them to hire the right people and pay them the right amount for their skill set. But surprisingly few tech companies certify the so-called "soft skills" of vertical industry knowledge or customer service skills or sales skills. This will have to change. In consumer markets, this can already be seen as HP hires support staff with expertise in digital photography, Intuit hires people with bookkeeping or tax experience, and Autodesk hires CAD experts. What we need is certification cubed: the technology, the soft skills, and the application in the customer domain.

But the cold, hard question about change starts at the top. Who will lead the converged VAS organization of the future? Is it the manufacturing mentality of the customer service executive? The consulting mentality of the professional services executive? The sales mentality of the geo executive? A seasoned general manager from the product side of the house? As you can imagine, there are pros and cons to each.

Lately we have also been mulling over this intriguing dilemma: Should we view service in the same organizational way we view products? What if there was a services R&D function, a services product marketing function, a services manufacturing function? These are all the same functions we fund for the product side. There is not much doubt that these are all needed. But it begs the tough question about how we might look in five years. Will we build parallel universes within our companies (one for product and one for service), or will our existing organizations learn how to perform the same functions for services as they do for products today? The latter may be unattainable for some companies—the service and product skill sets may simply be too different—but it is certainly how customers would prefer it. It is, after all, how they consume your offerings—together.

A DIFFERENT APPROACH TO SELLING SERVICE … AND MAYBE MORE!

Another organization that is likely to see huge changes during the era of services convergence and value added services is sales. From consumer technology to mission-critical applications, the job of selling services is becoming more complex, more costly and has taken on a larger role in—sometimes even leading—the overall customer sales process.

The complexity avalanche is already guilty of driving up the overall cost of sales for technology product companies. Your sales team already needs experts in sales, the customer's industry, and your products. Often we build entire organizations of pre-sales support personnel to supplement the limited product expertise of the sales force. Add to this the growing demand for customer "solutions" (products + services), and what do you get? An increasingly complex sales task. Given the industry's declining product margins and growing price pressures, this is not good news.

It is easiest to see the expanding role of service sales in consumer markets. At most large tech retailers today you don't just grab the product and check out. You find yourself making a trip to the services counter to be briefed on the full range of extended warranties, in-home service, and installation options that are now available for your product. When you contact the HP consumer support hotline, they may now try to up-sell you other related services, long-term contracts, or product upgrades. Every newspaper circular or direct mail catalog for consumer technology products today dedicates real estate to available service offerings. Even the product ordering process—either online at places like Dell.com or by phone—forces the customer through the services selection process during checkout. Services sales have definitely become more present and more sophisticated in the consumer and SMB technology industry.

In enterprise technology, dedicated service sales are nothing new, but the costs of selling service are skyrocketing. The complexity avalanche has increased the number and nature of services dramatically, and we are just getting started. Being a knowledgeable salesperson in all services across all products is becoming more and more difficult. As a result, many service sales today are requiring more than one service salesperson. There is an old joke in tech that the products have become so complicated that making a sales call requires a bus to carry all the pre-sales product experts. Now you are starting to need a second bus to carry all the service experts.

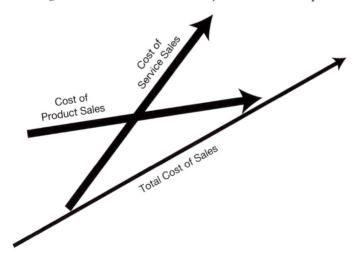

FIGURE 5.10 Effect of the Complexity Avalanche on the Cost of Sales.

At TSIA we believe the tech industry will soon see these three sales trends:

1. Services sales costs will increase faster than product sales costs.
2. Service sales cycles will shift from periodic to milestone-based, with each phase designed to progress a customer along a continuum of value achievement benchmarks.
3. Service sales will begin to pull through (sell) more products to existing customers than ever before. For many customers, the services sales team will become their primary advisor for both service and product sales.

For companies already into the services-led selling world, these trends have been around for quite some time. But for most tech companies—the ones with the product-led mentality and product-centric sales force—these trends are a major shift.

Can you imagine a *truly* services-led technology world? Customers would pay you an ongoing fee to achieve a constant escalation of value—moving from Point A (where they are now) to Point B (the next stage of success) around your area of technology. It would almost be like a ship moving through a canal using a series of locks, each phase building on the one before. We would move customers through stages of success, each phase predicated on the last. In this process we could not only deliver tiers of services, but tiers of features. Essentially we could offer an "atomized" product where customers started off with only the most basic features turned on. Then, as their consumption capacity grew through VAS, more product features would be turned on. Much like the gaming industry does today as users progress from beginner to intermediate to advanced. And each level of achievement could bring not only more value to the customer, but a higher level of revenue to us. Plus, as Don Norman from Chapter 1 said: "Unused features are not just neutral, they are negative. They clutter the product and make it more confusing for the user to succeed on their basic needs." As long as we staged the customer's success, moved them from "lock-to-lock" within the product, and continued to add new features, there would be a better foundation to ask the customer to constantly pay for more and more value realization. And we mean actual value, not potential value—a true "pay-for-performance" model. Plus, their emotional attachment to the product would grow as they "mastered" the product again and again. Every year, they would show up at your user conference with more and more medals on their chest. How much would they love that at a personal level?

Here are some more TSIA predictions about the future of enterprise services sales:

- Customer success benchmarks—and assessing customers against them—will form the basis of service sales discussions. Customers will want an objective assessment of their actual value realization and a systematic approach to how they can improve it.

- Services sales will be milestone-based instead of periodic as customers become willing to pay for VAS "mini-projects" that close key consumption gaps. As one VAS phase ends, another higher-level phase of VAS will need to be sold. These "projects" could last three days, or they could last 13 months. Thus service sales will become diagnostic- and prescription-based; consulting with customers on their status and capabilities as well as the current technology landscape and giving them a full understanding of what's possible with proper utilization. It will require a very high level of technical and application expertise, as well as knowledge of the customer's business, on the part of the service sales team.

- Service solutions will be far more customized and will pull service "parts" from across the converged services portfolio (PS, TS, FS, Ed, etc.) to meet the needs of a single customer.

- Services sales will slowly become the primary customer sales relationship and will expand its prescriptive solution to include products.

These predictions might be controversial at some companies. For one, many of your executives won't like the idea that the predictability of the maintenance revenue stream could be—at least in part—replaced by a more variable, project-based set of services. At first even customers, who like predictable costs, may not like the idea. But this will give way to the power of the approach and its business results. Another prediction that may be hard for some executives to swallow is that service sales could become the trusted advisor to the customer. And with that status will begin to assume the responsibility for selling products. These skeptical executives aren't the only ones. In fact, many people in service believe that service people shouldn't sell. Their notion is that service personnel should be the customer's true advocate and not be tasked with (or compensated for) disguising a selling objective as customer service. There is a clear separation, they maintain, between the selling motion and the service motion.

At TSIA we beg to differ. In the VAS era, ongoing diagnosis and prescription will be the key to the customer achieving the full benefit of the product's value, and those two actions must move as one.

Much like a doctor, service delivery organizations have a unique view into the customer's world. They know the problems, the holes, the adoption roadblocks, the technical limitations. By combining this understanding with strong knowledge of what is possible through the products and the VAS steps it takes to get there, the service organization is uniquely positioned to assess current performance, understand next steps, and prescribe the combination of services and products needed to get customers to a higher level of value achievement. Much like a doctor routinely combines diagnostic knowledge with an understanding of the latest medicines or treatments, services delivery staff must engage regularly with the customer with the intent of realizing the full value the product can provide. This can be seen as a "get healthy" program. Do we criticize the doctor for recommending an additional, fee-based course of therapy or a particular prescription? Hardly—even when he or she prescribes a more expensive therapy over a less expensive one because it may help the patient achieve superior results. So we believe it will be in the VAS world. Selling a product to a customer can be good for the customer if your goal is to make him or her as successful as possible. Service organizations are about to become an integrated part of the company's sales process.

One more reason to do this, we assert, is because there are more (admittedly well-camouflaged) sales opportunities that flow through the global services organization in a day than the inbound leads that a sales team sees in a month. Unfortunately, most of these sales opportunities fly right by because companies don't have the culture, the process, the training, the compensation, or the systems in their service organizations to identify them. In the act of providing service, many of our staff notice a customer's need for something more—more service, another module, a more current version of the product, more training—it's all briefly in focus. Unfortunately those opportunities usually disappear as fast as they appear in today's separate sales vs. service mentality.

The old transaction sales model is becoming more outmoded all the time. In the world of the complexity avalanche it is about the journey and the relationship it takes to navigate it. It is no longer just a transaction. As smart suppliers we must be constantly engaging in this diagnostic and prescriptive service/sales motion ... especially as the product commoditizes. It's not just to benefit us but also to benefit the customer. Each service opportunity is a chance to diagnose and prescribe. Each lost opportunity is a disservice to the customer.

Thus, a critical element in the profitability of VAS is going to be determined by the cost of sales. Here again we will see a difference between enterprise and consumer markets.

Enterprise tech companies may be able to more quickly see the positive total impact of their VAS efforts on their P&L or in their customer profitability models. That will encourage them to spend marketing and

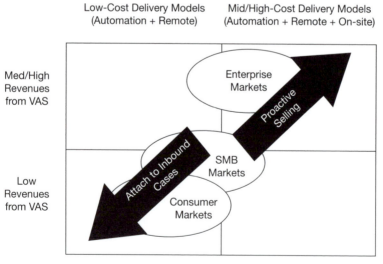

FIGURE 5.11

sales dollars on promoting VAS to their customers through outbound sales efforts.

As many consumer and SMB technology companies have already figured out, the way to start the VAS sales process may be to cross-sell it during their inbound support process (e-mail, Web, chat, or phone). Imagine a consumer tech support call coming into an offshore call center. While the rep has the customer on the phone, they engage in an extra conversation:

> *Rep:* "Sir, if I may ask, what are the primary uses of this computer?"
>
> *Customer:* "Well, it is our family computer and we bought it mainly for photographs and creating albums from them."
>
> *Rep:* "Is that going well?"
>
> *Customer:* "Actually, we're not using it much. My wife and son gave it a try but got frustrated trying to learn the software."
>
> *Rep:* "Sir, you might not know this but we have photography experts on staff that specialize in teaching photo applications to people over the phone. We could set up an hour-long appointment at a time that is good for your wife and son and our expert can remotely walk them through the process using your computer. We can even help build your first album using your own photos. It is only $69 for the hour and can make all the difference in making them successful with the computer. If either of them wants more training, they can schedule additional time to help with other applications."

So in this short discussion—less than 60 seconds of additional talk time—the rep has helpfully pitched a VAS offering to a highly qualified prospect with no incremental cost of sales except the talk time. Depending on where in the world the "expert" is, the $69 may be enough to pay for the tutoring. During that hour, your representative and your customer will accomplish amazingly positive things for you. You will design this VAS business process to:

- Register the customer in your database.
- Gently upsell/cross-sell additional products and services.

- Result in MUCH higher levels of customer satisfaction (very likely to lead to brand loyalty on future purchases).

- Create a vocal advocate for your products and your company.

- Gather critical information about product configurations, usage problems, etc., which can direct future product development.

- Gain the right to become your customer's trusted advisor.

Ultimately this relationship could expand, as someone or some company will need to become the consumer's IT department, proactively managing the technology stack and end-user adoption. Whoever assumes that role will definitely become that household's trusted advisor.

A similar conversation could also take place during a field service call for an office products company, as an example. The field rep may sell some of his or her own time to train new employees on your technology at a small business customer. They may sell four, hour-long sessions for one of your experts in a remote call center. Either way, it turns every service call into a sales opportunity.

This cultural shift toward playing a more integrated role in the sales process will be tough, but it is critical if today's technology service organization is to move to the next level. Does this mean that every service technician must carry an order pad and fill a sales quota? Well, it happens every day in banking or cellular service. When you call the customer service line, that representative may explain to you your options for higher levels of service or tell you about additional products and supplies that may be of interest. In some cases the representative may actually take your order. In more complex situations the service delivery organization may just pass opportunities off to another team. It is a fact that the complexity avalanche will cause tech companies to become even more heavily matrixed organizations. A VAS rep or a sales rep may need to pull in expert assistance from a lot of different places inside the company.

Remember, the goal of service is shifting. It is no longer to simply achieve and maximize product availability but to accelerate the consumption of product value—and it doesn't really matter how you get there.

Services need to learn that selling does not mean you have switched over to the "Dark Side." The right question for every person in your organization to ask is, "How can we accelerate the consumption of product value for this customer?" Call it selling if you must, but it is much more than that. It is getting your customer the best possible results by applying your company's people and product resources more productively. Think of Bob at the Geek Squad. He may be in the best position of anyone to help his customer get the most value out of her $1,500 investment in home theater equipment. If that means the customer has to spend another $170 in services or product, doesn't Bob owe it to the customer to explain the options and to consider her needs? He could even sell on the spot! Soon after the service event, the opportunity to make that sale will likely dissipate. Bob won't record it, no one else will follow up with it, and Bob will move on. The customer may never ask for help again. And for years to come, that consumer may get only a fraction of the value the product was capable of delivering had Bob taken the time to present a diagnosis and a prescription for more products or services. If the customer says no, so be it. But our job in services is taking a giant leap forward. Think of what that means to your processes, systems and your compensation models. We now owe our best thinking about value consumption to the customer every day. We must go on offense. The question is: Do we have the right people to do the new job?

The important point is this: The service motion and the sales motion will become one integrated process.

EFFECTS ON MARKETING AND PRODUCT DEVELOPMENT

One issue that has been hotly debated for years is why, where, and how services marketing is conducted. The story here is a classic one—and one we hope will be avoided in the case of VAS. For most tech companies in the 1960s, 1970s, and 1980s, there really was no such thing as services marketing. Customers were locked into a service vendor based on their product decisions; they had to have certain installation and maintenance services, and they bought them from the same company who made the product. What needed marketing here? Then in the 1990s, things changed. We talked about these factors in Chapter 3: third-party maintenance providers hit the scene and began to threaten the service market share of hardware OEMs; third-party professional services companies moved beyond government contracting or financial accounting to begin to take professional services deals away from software OEMs; corporate customers started to negotiate stricter SLA (service level agreements) and price concessions in the software contracts; even consumer companies began to price and package services on a mass-market basis. All these developments had one thing in common: Each meant that we had to actually start *thinking* about our service offerings. What were our target markets and which offer best suited each one? Who was our competition? How did the 5 P's (product, price, place, promotion, people) apply to services?

Usually, the services executive teams were the ones who first noticed the need for a services marketing function. And most of them did the right thing. They appealed to their marketing departments to do the job. But in many cases, marketing wasn't initially interested in services. Marketing was about products. This is why, to this day, the overwhelming number of services marketing organizations are within the service business, not the marketing department. This is a big mistake—but fortunately one that many companies are now figuring out. They are realizing that, more and more every day, the products and the services wrapped around them are becoming linked. Just like the sales process, they are not two separate offerings to be marketed in serial order but are all part of a single, total offering that gets the customer the highest possible level of value. So in our view, services marketing belongs in corporate marketing *as long as it gets the proper attention*. If corporate marketing is not serious about the staffing, funding, and strategic role of services marketing then let it stay in services. While it is not its optimum home, it is better than being a second-class citizen somewhere else.

As we mentioned earlier, another organization that will be greatly impacted by VAS will be the product development organization (R&D). There are two potentially big changes here. One is that VAS represents yet one more demand on limited development resources. We all know the classically tough decisions your company already faces here. Do we prioritize feature richness or time-to-market? What about either of those versus product quality? With a finite budget of people and money, development executives must prioritize. VAS is one more competitor for these precious resources. VAS requires that a certain degree of product metering capability is added to the product. We need to have built-in tools that aggregate and learn from incoming usage data, construct models of usage failures (and successes), and pass the information to those service design experts who will be creating the VM BOMs needed to roll out solutions to the customer base. These additions to the development slate are far from incidental. They will require programming hours, the imbedding of third-party tools, and perhaps additional computing resources. The business case for VAS must be able to hold its own ground against the pressure to get products out the door first, or it must present a business case as to why more money in the development phase is a powerful investment to VAS-enable the product.

The second—and perhaps more far-reaching goal—would be to move toward the atomization of products at the feature level. New users would start with only basic features turned on. As they succeed and are ready for more, we turn them on—much like today's video gaming industry does. In each phase, the customer might be paying us for the VAS services that lead them to success, and they may pay us more for new features as we turn them on. This would help customers in numerous ways by uncluttering the user interface, letting them "go deep" only in the areas that are truly needed, and letting them pay for actual usage and no more. It really makes great sense in an SaaS environment. It is also consistent with the direction that industry analyst firms like Forrester are advocating. In their 2009 Enterprise Software Licensee's Bill of Rights, analyst Ray Wang insists that only paying for the software in use by a customer is a fundamental right.[7] While he is specifically referring to unused licenses, you could imagine a day when customers want their software providers to stop charging them evermore money for evermore features that they are not using. "Pay for use" is a concept that is likely to stay around for awhile. It will make the economic returns on VAS even more immediate and essential.

WHO IS GOING TO PAY FOR ALL THIS?

We realize that the bottom line of your decision to launch VAS is just that: What is the effect on the bottom line? So who is going to pay for this? Are there new service revenues associated with VAS?

Let's start with paying for the costs of VAS. As we discussed in Chapter 2, the gross margin improvements of the enterprise support and maintenance business over the past decade have gone to fund many other activities. They have financed increasingly high product sales costs, shrinking product margins, frequent professional service losses ... basically allowing many companies to hold profitability flat against a lot of weakening fundamentals. And realistically, short-term financial challenges (making the current quarter and year) will not allow them to write a huge check to create their VAS offering all at once. It will start small and grow.

Well, we are heading back to the well again to fund the transition to VAS. In essence we are going to fund it by increasing our organizational commitment to driving down the cost of all service across the silos. But this time it's different: We will redirect a significant portion of the resulting savings into building VAS. To make this transition to VAS, service organizations will need some serious help. They can't just "pedal harder." They have been doing that now for at least five years and the returns on current productivity initiatives are declining.

This approach will require everyone to pitch in. The development organization has to better engineer the supportability of the product (the cheapest single way to reduce service costs). The professional service teams must not only continue to improve margins to contribute to the VAS investment pool, but also make a commitment to leaving their customers in a fully implemented, fully deployed, highly supportable state, and nothing less. Since customer service is the ultimate back stop for every customer problem, their costs are directly linked to product quality and professional service engagement results. Whatever work is left to be done from an underperforming offering leads directly to increased customer service costs.

There is a certain symmetry to hanging this on the services organizations to self-fund the transition. They believe in VAS, they will manage much of its evolution, and they are the best equipped of any organization to make it a reality. But the CEO and CFO have to be willing to let them keep their savings—at least a significant part of them. If you keep robbing their piggy bank to pay for other departments' problems, you will never get to VAS, never solve the complexity overhang for your customers, and markedly delay your next product refresh cycle.

The second question is whether you can supplement these cost-reduction and reinvestment efforts with new revenues. In our survey of IT executives, the overwhelming majority said their customers were willing to pay for services that enable more effective usage of enterprise technology. Twenty percent of respondents reported that paying for those services is a "must have." That is much higher than the percentage of customers that indicated no interest in such services. Naturally the vast majority indicated a willingness to pay depending on scope and complexity—not a surprising position to take. But overall, nearly 90% of these enterprise IT customers expressed a willingness to purchase VAS-type services.

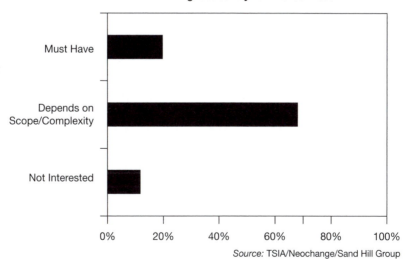

Willingness to Pay for VAS Services

Source: TSIA/Neochange/Sand Hill Group

FIGURE 5.12 Buyers Are Willing to Pay for Additional Services That Enable Effective Usage.

So what are the options for revenue increases due to VAS? Each company must make its own decision about how VAS services fit in their overall offering. There are at least four main options:

1. Create a separate VAS subscription annuity program that sells for a separate fee.

2. Bundle VAS services into your premium maintenance offering to increase its adoption rate.

3. Bundle VAS services into your standard maintenance and support offerings to increase renewal rates and reduce discounting pressure.

4. Create a new à la carte VAS project menu that sells individual projects at incremental prices.

For many enterprise tech companies, selecting the right strategy begins with an honest assessment of the health of your maintenance business. Due to the incredible economic importance of that revenue stream, strengthening its value proposition might be job number one. For consumer companies who do not sell much in the way of maintenance and support, creating a menu of à la carte VAS services will most likely be the way to go.

If your company uses channel partners to deliver a significant portion of your sales and/or service function, you may need to work out what the channel is capable of providing in the way of VAS services versus what you must provide on an OEM-direct basis. Many of your partners will be delighted to get into the VAS business as a source of growth that continues on through strong and weak product cycles, but they will need your company to VAS-enable the products, develop the VAS BOMs, maintain the product usage best practice models, etc.

The key to all these changes is going to be the strength of your company commitment. If you position VAS internally and externally as a central strategy of your company, a reason why customers should buy from you, and you supplement that with a sales process and comp plan that embodies the elements in this book, you will drive both new product and service revenue. Ultimately, delivery may turn out to be the least tricky part of the equation.

So, like any other strategic initiative in your company, it all starts with belief at the top. The customers and the employees will follow. It is a big change—one that will be hard to manage. But look around you—isn't the handwriting on the wall? It is time for tech companies to take responsibility for driving customer adoption and value.

6 | Courage to Chart a New Course

PUT SIMPLY, IT'S ABOUT GROWTH—YOUR COMPANY'S GROWTH. Technology products in market after market suffer from a growing consumption gap. Left unchecked, this gap will slow the sale of products. So let's roll out the compelling reasons why the time has finally come to become a truly integrated product AND service business.

1. **Companies that have a well-developed and integrated services capability will gain market share over those who don't.** Customers are favoring companies that provide a total solution and can be counted on to make them successful. Your customers are demanding more than a bag of bolts. They want you to deliver a complete solution and maximum value. That means services must be a centerpiece of your offering. Even your existing customers—many of whom you thought were locked in—are increasingly vulnerable to the siren's song of new services.

2. **The consumption gap threatens to slow repeat sales to existing customers.** If customers can't use the capabilities they have already paid for, then why pay for new ones? Whether it is resource-constrained IT departments or patience-constrained consumers, the complexities of process, technology, and features are gating the sales of products today.

3. **Services can make your addressable market larger by enabling less technically sophisticated customers to be**

successful with your products. If all you do is sell to early adopters you may be missing 60% or more of your potential customers.

4. **Key new features are in jeopardy of having minimal market effect due to customers' inability to use them.** Don't put your best features on the complexity "top shelf" where very few customers can actually get at them. The value of your precious R&D investments could perish before your eyes.

5. **Services can open new markets and extend the life of products.** What services will be needed in cloud computing? What about for the emerging market for home health care equipment? Name an exciting market and you will find a need for VAS standing nearby.

6. **Services sell regardless of the economy.** The product business has been seriously affected by the economic downturn of 2009, but services revenues have proven much more resilient. We have conclusive data to prove that tech companies are much more likely to be successful over the long term if they build service businesses that contribute in a steady and dependable manner to the financial performance of the company.

7. **Maintenance—as a value proposition—is in real trouble.** Failure to replace that highly profitable revenue stream with a new one could be devastating to hundreds of enterprise technology companies. Will your company bury its head in the sand and hope the pressure subsides or will you take action now? Remember, moving to VAS will take time.

8. **Services can be a great revenue and margin opportunity.** It's not that customers won't pay for service; they still see the need to pay and receive THE RIGHT services. The complexity avalanche has begun to bury them. More and more they will pay for VAS services that truly advance the value of your product. This is especially true if the result is increased end-use adoption, which is the key to real value creation.

So if this is all true—or even mostly true—for most tech companies, then what's holding us back from truly adopting service as a strategic lever to grow the product business? We at TSIA would submit that there are four key boundary conditions that have kept us in a services "lock down" for 30 years, despite clear messages from the marketplace:

1. Overcoming our product DNA.
2. Wall street versus the customer.
3. The organizational "box" driven by current financial reporting practices.
4. We can't calculate the full economic impact of services.

OVERCOMING OUR PRODUCT DNA

Let's face it: Tech has always been a product business. It was started by product people and it is still run by product people today. Companies from Adobe to Yahoo! created billions in shareholder value by following the well-proven notion that features dominance is the most predictable driver of market dominance. Importantly, this was also the part of the business they know best. The service business is, and too often remains, somewhat foreign to many of their native knowledge and skill sets. Sure, there are a few tried-and-true service strategies (like, it is good to build a maintenance business if you can). But to truly integrate services into your corporate strategy you must push yourself (and your investment dollars) into areas that are perhaps not as well understood and where new business models are rapidly unfolding. Until TSIA came along, services were an area of the tech market loaded with myths and not much hard data or valid performance standards. This can lead to a lot of internal debate about how best to think about and run services.

We think there are good answers now, and we can help companies to find them. But until the product executives become more comfortable in the services business, they are going to remain slow to integrate it into their strategic thinking and their investments. IBM has done it. HP is on the way. So are Oracle, CA, and others.

In consumer technology Apple has shown us that, despite great product designs and simple user interfaces, there is still a large consumption gap. It is stealing market share and maintaining margin advantages even in cutthroat businesses like notebook computing, in part by making value added services easy to access from any Apple store. The reality is this: Virtually every technology product in every market has a consumption gap. With the complexity avalanche gaining size and momentum, tech companies must rapidly begin to integrate value added service offerings that drive end-user adoption into their corporate strategy.

WALL STREET VS. THE CUSTOMER

Ah yes … what will Wall Street say? Well, until recently, it truly would have been like throwing a boat anchor around your stock price to tell financial analysts that you were going to increase your investment and commitment to services. (Again, we mean true service, not product-as-service). They would tell you that services don't scale, that professional services are a margin drag, that services will add headcount and slow the rate of product innovation. "These guys must lack a winning product vision if they are doing THIS!" might be a reaction by an analyst rooted in the product-focused history of the past couple of decades.

Maybe … just maybe … things are starting to change.

First, there is a growing recognition of the value of a strong and healthy maintenance business. This is particularly true in enterprise software and hardware markets. Analysts recognize that this is a beautiful business: recurring revenue, strong cash generation, high gross margins, and low incremental cost to serve. Especially in the recently tough economy, Wall Street is more curious and more rewarding of this one service line than it has ever been. It's not that Wall Street loves the service dimension; it just loves the economic model. But it is a start.

Ironically, those analysts who have begun to love services such as maintenance have an almost equally compelling fear about the margin drag represented by project-oriented professional services, as we have pointed out. Software companies especially fear that rapid growth in low-margin, project-based services like professional services will drag down the overall margins of the company.

If your company is small but growing or if your company has always pointed to your channel partners to act as your services ecosystem, then the decision to invest in the scale necessary to be profitable in some service businesses is a tricky one. But, as we have pointed out, we believe that NOT expanding your service capability somehow—either directly or through partners—is no longer an option. Your customers need the services and their absence will confine the growth of your product business. So the question is not if, but how.

Whether your customer is a consumer, small business, or global multinational there is a way to scale service revenue and margins profitably. And even if not every service business is able to generate a 50% margin, a dollar of gross margin is a good thing for shareholders, not a bad one. We should be more focused on those dollars than on that percentage.

A second reason why Wall Street may be ready to rally around the Rise of Services is that those companies with a balanced mix of products and services are outperforming product-only companies in the current economy. The IBM model (greater than 50% of revenue from services) has proven itself after a rocky start, and others that have followed this pattern are being rewarded as their better-diversified businesses continue to produce profit growth through good economic times and bad. Figure 6.1 shows the profitability of Service 50 companies grouped by their concentration of service revenues.[1]

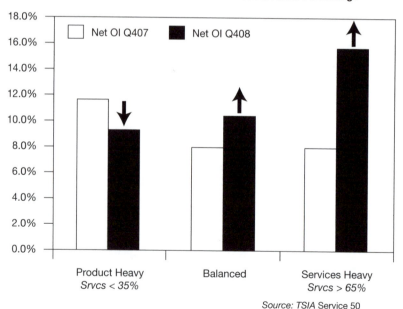

Service 50 NOI Based on Service Revenue Percentage

Source: TSIA Service 50

FIGURE 6.1 Revenue Mix and Profitability: Services as a Drag? Not!

The group of companies with less than 35% of total revenue coming from services saw their operating income decline in the fourth calendar quarter of 2008 versus 2007. Companies with 35% to 65% of revenue coming from services were more profitable, despite the significant economic challenges in late 2008. The companies with a heavy services-oriented revenue mix (less than 65%) had the strongest profit performance and the greatest year-over-year increase as a group. Let's face it, services is an attractive financial safety net—one that every product company should be trying to build.

To these two solid starting points, we would add some additional thoughts for financial analysts and the business media to consider.

One is that Wall Street accepts that the Rise of Services—even well past a 50/50 revenue blend—is simply an inevitable part of a global macroeconomic trend toward services, not a corporate decision that management has the option to make one way or the other.

The rise of services is everywhere, not just in tech markets.

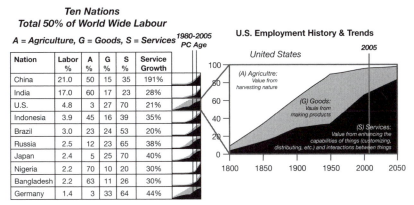

Ten Nations
Total 50% of World Wide Labour
A = Agriculture, G = Goods, S = Services

Nation	Labor %	A %	G %	S %	Service Growth
China	21.0	50	15	35	191%
India	17.0	60	17	23	28%
U.S.	4.8	3	27	70	21%
Indonesia	3.9	45	16	39	35%
Brazil	3.0	23	24	53	20%
Russia	2.5	12	23	65	38%
Japan	2.4	5	25	70	40%
Nigeria	2.2	70	10	20	30%
Bangladesh	2.2	63	11	26	30%
Germany	1.4	3	33	64	44%

International Labor Organization

The largest labor force migration in human history is underway, driven by global communications, business and technology growth, urbanization and low cost labor

Source: International Labour Organization

FIGURE 6.2

According to the International Labour Organization, in 2006, for the first time in human history, agriculture was relegated to the second largest global employment category. Today the service sector's share of global employment stands at over 40%. This general trend can be seen in country after country. We are becoming a service economy.[2]

As we have already pointed out, the tech business is demonstrating the same tendencies.[3]

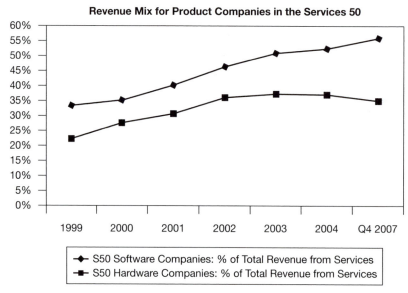

Revenue Mix for Product Companies in the Services 50

Source: TSIA Service 50

FIGURE 6.3 Shift in Product: Service Mix.

Services revenue is gaining on product revenue in tech market after tech market. And the shift in profit mix is even more dramatic. For some tech product companies today ALL the profit already comes from services. The total cost to design, build, market, and sell products for these companies is simply greater than the revenue they can derive from the customer. But the service and supplies aftermath of these products enables an overall return for the effort. The product has become the razor and the service has become the blades.

So if you can see this happening globally and you can see it happening in tech markets, is it really something companies can or should avoid? Probably not. Of course, what Wall Street fears is that the growth of services (even maintenance) as a percent of total revenues means that the growth of the product business has declined and will trigger a reduction in total revenue growth, EPS, and free cash flow. Once we link services to product sales more clearly and help Wall Street understand that the growth of services revenue is natural and healthy, it will be seen as less of a negative.

No one in the world of tech understands, accepts, or even embraces the inevitable rise of services like IBM.

IBM stands alone among product companies in its investment and its return on services. Let's take its commitment to the emerging area of services research and development. IBM Research has over 3,000 scientists working on fundamental innovation in many fields. Up until 2004 services was not really one of them. In fact, in 2005 the entire service research function consisted of only 50 scientists. Things have changed. Today IBM Research has nearly 500 scientists focused on services innovation. They are producing returns from their innovations that are measured in the billions of dollars. They are the fastest-growing sector of IBM's total research budget and they are getting further ahead of their competition every day. They are getting a total ROI of between 8 and 10 times on their service R&D investments.[4] And they are realizing them rapidly.

What did IBM see that other companies did not? We would say three things:

1. It recognized the absurdity in the ratio of R&D spend as a percent of revenue between products and services. Product revenue had dropped below 50% of total revenue but received nearly 100% of the research budget.

2. Innovation in services is possible and will respond to traditional innovation practices. With the rapid proliferation of technology that is being applied to service ecosystems today and the growing understanding of the science behind service, there are better service mousetraps that can and are being built. IBM knows that service processes can in fact be improved, scaled, and made increasingly asset intensive.

3. Services will contribute mightily to future growth, profits, and differentiation. IBM sees that product categories left and right are commoditizing—that as computing moves into the cloud, there will be fewer units of servers and operating systems sold; that the

game of the future is around software and that software and services are forever joined at the hip. It knows that the key growth game in tech will soon become the ability to apply services to customers that transform processes and practices; identify, design, and create new business opportunities, or otherwise advance the value of technology for the customer. Soon services will be the most important enabling ability in tech.

It will be up to you to decide what the implications of these trends are to you and your business. As we have said before, we at TSIA believe the right approach is to allow your current services business to reinvest a portion of its savings from ongoing operational efficiency improvements into services innovation and capability, that is, to stop robbing the services piggy bank.

But what is clear is that it IS the Rise of Services in tech today. Wall Street analysts who discourage your investment in services may be helping your transient investors this quarter or the next but they are doing a massive disservice to your long-term shareholders and, more importantly, to your customers.

THE ORGANIZATIONAL "BOX" DRIVEN BY CURRENT FINANCIAL REPORTING PRACTICES

Peter Drucker taught us all that "what gets measured gets managed."[5] Unfortunately, we at TSIA have made the following subsequent observation: "The way it gets measured is the way it gets managed." Nowhere can this be seen more clearly than in the way that tech companies measure and manage their services businesses.

Classic accounting style, classic management style, classic compensation theory, you name it—they all embrace the concept that measuring and holding people accountable for bottom-line results is the best way to manage. Who can argue? It works. But what if one division's bottom line had an effect—positive or negative—on another? And what if the company lacked the visibility to establish and track the connection? What if the drive to cut costs and improve profitability in a single division had the effect of reducing the overall company's growth and profits? Welcome to the frustrating world of financial reporting for service executives.

In our experience with virtually every major tech company in the world, we have yet to see a reporting system (and a culture) that does not behave like this:

Dividing the Business into Divisional Stovepipes

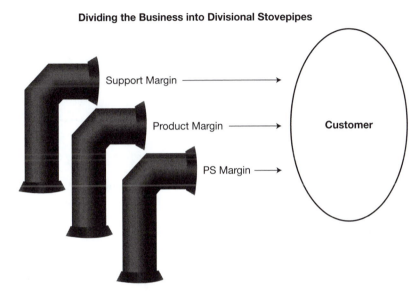

Support Margin ⟶

Product Margin ⟶   **Customer**

PS Margin ⟶

FIGURE 6.4 Organizational Stovepipes.

In our example, the company views and manages the company as three divisional stovepipes: the product business, the support/maintenance business, and the professional services business. Now, your company may have slightly different stovepipes. Maybe education is a material business, or outsourcing, or supplies. And certainly you will have multiple product stovepipes to reflect your different product lines or markets. But you get the picture.

At TSIA we have observed a remarkable consistency from company to company in how these divisions:

- Are financially reported—each as a separate P&L.
- Measure and compensate their executives—on the margin performance of their own stovepipe.
- Get managed—as largely individual businesses with a subservient consideration for or about other divisions.

So the management of the company using this divisional mentality—driven at its heart by current financial reporting practices—might think like this:

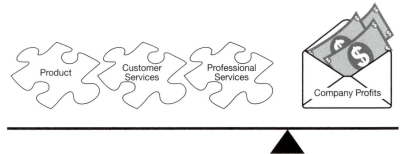

FIGURE 6.5 1+1+1=3.

The company is almost three different companies. We would argue that it is also what it feels like for customers who do business with this company. "I have my product salesperson, my support salesperson, and my professional services salesperson. While one of them may claim to be my overall account manager, it is clear that each is a separate organization who only communicates with the others in the event of a problem." For new customers, this may initially be hidden by the frenetic courting activity known as the initial sales and install process. But established customers who have done business with the company for a period of time know this to be true and become more and more frustrated every day.

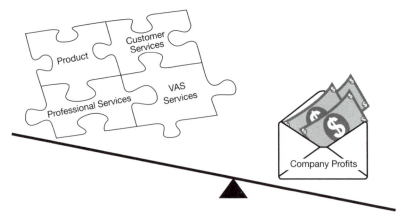

FIGURE 6.6 Tomorrow's Success.

We would argue that customers want to see the company as an integrated stack, not three silos. Information, processes, costs, and activity can and should regularly flow up and down the stack to ascertain and deliver the best total outcome for the customer. As we have already outlined, this converged approach will lead not only to better customer outcomes but also to improved efficiencies and lower costs across the silos. Most importantly, we believe that this approach will maximize what matters most: the total profitability of each account.

Measuring account-level profitability is not easy. While many integrated financial application vendors maintain that their products can do it, we don't know many executives that are consistently getting this view. But we at TSIA think it is a critical addition to current reporting tools and will drive the right behaviors for both customers and shareholders. We also believe it can be done—that the data sets are readily available to companies but the companies do not consider the cost and hassle of compiling them to be a worthwhile investment. We think the inability to architect and track this critical financial view is one of the two great management failings of services executives. We should have gotten to this view earlier. It is time for this to change.

If our siloed financial reporting and management models make executives hold back services to maximize the profits of an individual silo, we will miss a huge opportunity for growth. Companies like Salesforce.com are learning that more service means more users. Since more users mean more monthly license fees, they are breaking free from the traditional financial reporting boundaries of service and doing VAS things like proactively tracking down poor and non-users and delivering services to get them up and running. Since Salesforce has only one revenue stream, the silos tend to go away and the whole company focuses on whatever set of activities drive that one number. The company, in effect, has only customer profitability to worry about and it has learned to make investments in service that others don't. It works for Salesforce and it can work for you.

Vonage is the same way. When I switched over to their phone service in early 2009, they must have called me two times in the first week to ask me when I was going to install it and then two more times to make sure it was all successful. How come you don't get the same calls from Palm or Sony? Because their model is about making money off the transaction. Vonage must have successful adoption or it gets nothing. Just this simple difference drives a fundamentally different service strategy.

We are not saying to relieve the cost pressure on service executives, or excuse them from having accountability for service profitability. We have been down that road before. We know that trick doesn't work. What we are saying is that the current siloed approach is preventing us from making the best possible decisions for customers and shareholders. We need a second view.

WE CAN'T CALCULATE THE FULL ECONOMIC IMPACT OF SERVICES

Calculating the likely financial impact of VAS is a challenge because it has the potential to affect so many parts of the business. This issue is related to account profitability but in a more comprehensive way. In the spring of 2009, before the Technology Services World conference, we convened over 20 top service executives from global tech companies to ask a simple question. "What are all the economic impacts on the company that could be driven by the availability, quality, and price of services?" The group identified well over a dozen key financial or operating measures that are likely to experience change.

- Drives consumption of features.
- Drives upgrades.
- Drives consumption of software licenses or equipment capacity.
- Drives consumption of supplies.
- Opens new markets for products.
- Generates service revenue dollars.
- Generates service margin dollars.
- Creates reference customers/improves overall customer satisfaction.
- Improves renewal rates.
- Increases total account profitability.
- Increases total account spend.
- Drives product feedback to R&D.
- Reduces cost of sales.
- Gains knowledge of the customer.
- Fixes products.

- Creates differentiation and solutions that result in market share gains.
- Extends the life of mature products.

You might even think of a couple more. But the point is this: Of the 17 contributions on this list, service businesses are usually measured on just three: service revenue, service margin, and customer satisfaction. The other 14? We believe them. We might even track a couple of them. But the reality is that the vast majority of economic impacts emanating from service availability and quality are not measured. That is a poor state of affairs (especially for software companies who boast advanced insight into corporate data as a cornerstone of their product). Why is this? There are four reasons, according to these executives:

1. We don't know what to measure.
2. We aren't collecting the data today.
3. There is a lack of executive interest.
4. Other departments will try to take credit if there is an improvement.

In other words, it's hard to do. So should we bother? Consider the following chart that maps the business and financial metrics that we at TSIA believe will move—either positively or negatively—based on a company's decision to implement VAS.

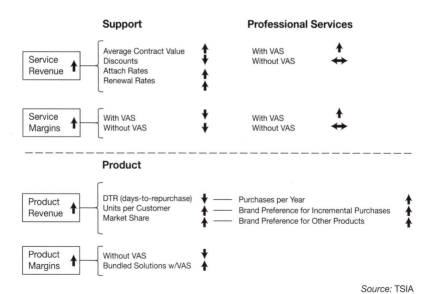

Source: TSIA

FIGURE 6.7 What Are the Financial Impacts of VAS?

- Support and Maintenance Revenues—with maintenance contract discounting becoming standard practice and rapidly commoditizing product categories combining to threaten support revenue growth, VAS may be the savior the industry desperately needs. By shifting the value proposition from high system availability to advancing product value and driving end-user adoption, companies will be better able to defend or even increase support prices, while simultaneously boosting attach and renewal rates. In the case of Oracle ACS, Oracle has created a support add-on offering that is generating over half a billion dollars per year of new revenues AND, because of its policy that all ACS customers must also purchase its premier level of support, boosting revenues for the traditional maintenance business.

- Support and Maintenance Margins—due to flattening revenue growth and increasing labor costs, support margins have already begun to plateau or decrease for many companies. This trend will likely accelerate. Without a doubt, VAS will create additional costs that could further reduce existing margins. The question is simply whether the revenue gains from VAS will more than offset the incremental delivery costs. The other hope is that if overall service costs benefit from services convergence initiatives, then some of the saving can be reinvested in VAS so that the company's total spend on services delivery does not increase.

- Professional Services Revenue—the existence of a VAS capability will open many new categories of professional service opportunities designed to increase end-user adoption (EUA) and overall product value. Due to their very high ROI, we believe that customers will be eager to utilize these services. The net effect of VAS will be more professional service projects and an increase in professional service revenue.

- Professional Services Margin—perhaps even more importantly than boosting professional service revenue, VAS offers the opportunity to significantly improve professional service margins. By developing a highly capable, centralized VAS organization, portions of professional service projects can be moved from a traditional on-site model to a far lower cost, centralized model. Learning to "chunk" professional service projects and allocating the chunks to the most effective and efficient global resource is the path to improved professional service margins. This will be done by separating front-end (at the customer site) versus back-end (in a central solution center) professional service activities. VAS is an important step. The VAS organization could be used not only by your direct professional services teams but also by your channel partners. They would see it as a real benefit to have a centralized VAS organization to provide components of projects faster, cheaper, and better than they could do alone.

But the really big win is found below the dotted line:

- Total product sales revenue to the existing customer base will increase because VAS will reduce the cycle time between major product repurchases (more in Chapter 7).

- Customers will buy more units because of the higher value they are receiving.

- Customers will shift share-of-wallet to your company (over your competitors) because your company delivers higher overall value and does it in less overall time.

- Product margins on repeat purchases will improve because customers have fewer choices (assuming your competitors don't offer VAS) and because they are delighted with the success of your efforts. In other words, they will push less for discounts.

- Cost of sales will be reduced for the same reasons.

- Your market will expand as less technically proficient customers feel comfortable that your VAS service offerings can make them successful. That gets them off the sidelines and into the market.

- Finally, your overall ASP (average selling price) will go up as you bundle more and more VAS services to create total solutions for your new and existing customers.

In summary, most companies have done a poor job of quantifying and measuring the relationship between these business results and investments in services. They have not settled on the non-service business metrics that they expect to move, captured the before-service and after-service snapshots of the customer, nor rolled that data into a credible model.

BREAKTHROUGHS IN SERVICE MEASUREMENTS

Measuring Total Economic Impact of Services

NEW MEASURES:	OLD MEASURES:
• License Consumption	• Service Revenue Growth
• Product Margins	• Service Gross Margin
• Cost of Sales	• Attach Rate
• Days-to-Repurchase	• Renewal Rate
• Market Size	• Service CSAT
• Product Adoption Rate	• Discounts
• Market Share	• Project Profitability
• Customer Insight for Product/ Solution Development	
• Services-Led Markets	

Because we cannot quantify and prove these linkages, CXOs think not about investment in services but only of cost reduction. This has negatively impacted customer success, satisfaction, and product sales outcomes. If we had built this case earlier (the other great management failure of service executives), the relatively poor levels of customer satisfaction that the tech industry is famous for just might have been avoided. It can be done. And it might be the biggest lever to drive product growth and improved financial performance that we DON'T have a handle on today.

7 | The Power of DTR (Days to Repurchase)

A NEW MEASURE OF CUSTOMER SUCCESS

Companies have been trying to measure and manage customer satisfaction for decades. They've conducted surveys, collected data, identified trends, created indices, written mission statements, developed loyalty programs, and spawned hundreds of books. The result is a series of customer satisfaction "wisdoms" that get spread around to all levels of the organization, such as:

- *As little as a 5% increase in retention can improve a company's bottom-line profitability between 25% and 85%, depending on the industry.*
  Fred Reicheld—*The Loyalty Effect*

- *Selling to a new customer often costs 5 to 10 times as much as selling to an existing one.* The Customer Service Institute

- *A very satisfied customer is nearly six times more likely to be loyal and to buy from you again or recommend your product than one who is just satisfied.*
  Thomas Jones and Earl Sassar—*Why Satisfied Customers Defect*

- *Each unhappy customer will share their grievance with at least nine other people, and that 13% of unhappy customers will tell 20 people or more.*
  U.S. White House Office of Consumer Affairs

While these may all be true—at least to some extent—what is missing is a reliable, usable metric that directly links your company's activity to desired customer activity and on through to increased profits. The industry believes that customer satisfaction accomplishes that goal, but no one has been able to produce a set of tools to map investments in customer success to the financial metrics of the business. When I ran Prognostics I faced this dilemma for a decade. We were the industry leader at measuring and analyzing customer satisfaction for high-tech companies. I think our approach to customer satisfaction measurement was the best at that time. It was used by hundreds of tech companies around the globe. But it still had weaknesses. They all do. In the back of my mind, I had this nagging feeling: I kept thinking that what companies were REALLY looking for was not a scorecard but a gas pedal!

As an industry, if we want to keep growing product revenues and margins from our increasingly important base of existing customers, we need a different approach, one that focuses on accelerating the customer's time-to-value in a manageable way. The best way to do that is to shift the focus from the abstract concept of satisfaction and what customers *think* to the concrete reality of what they actually do. After all, isn't that what we really care about? A satisfied customer who is not repurchasing is not much of an asset. It's hard to get that customer to improve your income statement or put cash on your balance sheet.

Fortunately, we at TSIA have come up with a new idea that has captured the imagination of a lot of people. We call it the Customer Base Revenue Index.

The concept can best be articulated by stealing a bit of Newton's second law: Momentum = Mass × Velocity. So what's our proposal? Simple: The right measure of the health of the customer base should be its ability to generate revenue. Customer Base Revenue Index (CBRI) is our version of Momentum. We all want the CBRI to accelerate. Quickly.

This, of course, raises another good question: What drives revenue from the existing customer base? Two things: how much customers spend on repurchases, and how often they do it. So we've replaced Newton's Mass × Velocity part of the equation with this: Average Sales Price (ASP) ÷ Days to Repurchase (DTR).

We'll be the first to admit that the ASP part of the calculation is hardly a breakthrough, and is already widely tracked in a variety of industries. We suggest, however, that you be very selective in what transactions you count in your analysis. Unless you're selling supplies, what you really care about is significant purchases (if you are selling supplies, you can use a measure of supply sales per month, per customer). Thus your ASP calculation might include only purchases made by existing customers for certain products or for a minimum dollar value. That's up to you. What's important is that the sales transactions you measure within your existing customer base are the sales transactions that you want. You might, in fact, want more than one metric—maybe one that tracks overall sales and another that focuses only on your most strategic product line.

The real breakthrough here is a simple calculation called DTR, for days to repurchase.

Before we get into the mechanics of DTR, let's take a quick look at another three-letter acronym that you've probably heard: DSO (days sales outstanding), which is a well-known measure of accounts receivables health. Basically, it's a calculation of the average number of days it takes you to get paid from the time you invoice a customer. Say your company does $120 million in sales per year, or roughly $10 million per month. If you have $20 million in receivables, that's 60 days' worth of revenue, or a DSO of 60. The longer it takes to get paid, the higher the number and the poorer the health of your accounts receivable. If DSO is up, that's bad. When it goes down, that's good.

DTR is similar, except that instead of measuring the time it takes customers to pay their bills, it measures how long it takes for your customers to make another purchase. DTR applies equally well to products, services, or both. Let's say that average DTR at your company is 500 days (about 17 months). If you were to take steps to decrease DTR to 400 days, and the ASP stayed the same, you would boost the sales revenue from your existing customer base by 25%! What would that do for your stock price?

So let's look at the whole CBR index equation. In a healthy organization, the CBR index variables would be moving this way:

⇧ CBRI (Customer Base Revenue Index) =

⇧ ASP (Average Selling Price) ÷ ⇩ DTR (Days to Repurchase)

We want the customer base revenue index (CBRI) to go UP. We will do that by driving the average selling price (ASP) UP and the days to repurchase (DTR) number DOWN.

So the first thing you will need to do is calculate your baseline CBRI, against which future progress will be measured. Let's take a simple example for an enterprise software company that has a current ASP for subsequent or follow-on transactions (more licenses, new modules, service solutions, etc.) of $1 million. Then let's say that the current average DTR for post-sale transactions is 500 days. The CBRI baseline for that company would be 2,000.

ASP	$\div$	DTR	=	CBRI
$1M	$\div$	500	=	2,000

To put a revenue value on a single point of CBRI we would simply take the total revenue of all these transactions—say $100 million—and divide by 2,000 to get a revenue dollar per CBRI of $50,000. Thus, every point of movement, up or down, in the CBRI is equal to $50,000 in post-sale revenue.

Again, to illustrate the power of DTR, let's say that there were no new products and no expanded sales capacity that could drive up the ASP. All we did was to focus the company on implementing a VAS strategy that drove DTR down to 400. The CBRI would then climb to 2,500—an increase of 25%. In real revenue per year, that would be an increase of $25 million. By simply running your average gross margin calculation against that incremental revenue, you begin to see the true and exact ROI of investments you make to drive down DTR. If a specific budget request of $2 million can drive the CBRI up by 100 points (using our example), it would drive $5 million in product revenue growth. If that was software at an 85% gross margin then we have a ROI of less than six months.

Using the CBRI should take a lot of the uncertainty out of some typical investment decisions. The risk of investing in customer satisfaction without generating a return is dramatically lessened. The CBRI is based on what your customers actually do with their wallets, not just how they are feeling. It will focus your company more rigorously on the entire post-sale process, what the steps are, what the internal systems and processes are for both you and your customers, and where the breakdowns occur that drive up DTR. This modeling process will show you every delay, and now you will know how many revenue dollars you are losing as a result. Eliminating the delays not only accelerates your customers' time-to-value, it increases their ROI and puts them back in the market for more products and services faster. And that is good for your shareholders.

ACCELERATING DTR THROUGH VAS

Will VAS reduce your company's DTR? Well, as we have pointed out, getting value from technology takes time for your customers, and their success is not ensured. Yet organizationally today, tech companies don't line up well with what happens along the customer's path to value.

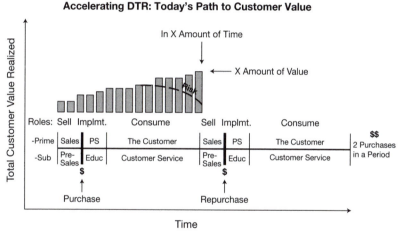

FIGURE 7.1 Accelerating DTR: Today's Path to Customer Value.

Before the purchase, your sales and pre-sales teams are there to help educate the customer about both the state-of-the-art in your product category as well as the specific features and benefits of your product offering. Sales assumes the PRIME role and pre-sales is the SUB role. The first milestone is actual customer purchase. Once the deal is signed, the PRIME role switches. In enterprise tech companies, the professional services team takes the lead. They have a specific project to get the customer implemented and deployed. Customer education teams are there to work on end-user training (SUB). Once the project work is complete, the customer begins to actually consume the value of the project. As we have pointed out throughout the book, the customer is now in the PRIME role. The time it takes to actually realize value and the amount of benefit they actually get is largely left to them to determine. Sure, the customer service teams are there to help maintain product availability (SUB), but at the end of the day, customers are mostly on their own. The amount of value they attain and the period it takes them to realize it can vary greatly. For the first time, the tech vendor has lost control of the process. In the case of consumers, the risk period extends all the way back to the point-of-purchase in most cases. The customer becomes their own PRIME as far back as the installation process.

I think we could assert that most customers do eventually get to a level of value that they are (at least somewhat) happy with. But the risk of failure or underachievement of expectations remains. And even more importantly, as we are pointing out, the length of time that takes has a direct bearing on the financial performance of both you and your customer. Yet neither the amount of elapsed time nor the degree of success is something that today's tech companies have control of. Worse, they don't usually even have real-time insight into the outcome for most of their customers.

Eventually, even in today's operating model, it does start to become clear about what some of the results were for a few of your customers. How? Because the successful ones go on to purchase more and they become big customers. And a few of the customers who failed miserably might get pissed off enough to call someone. But what remains your primary tool for gaining insight into the outcomes of the vast majority of customers? It's the customer satisfaction survey.

This is simply not an adequate strategy. Customer satisfaction surveys usually lag far behind the time when actual customer results are being determined. Even then, the customers are aggregated into a survey "sample" where their individual status and needs get lost in some "average score." It is simply too little, too late.

Not only is our lack of a proactive, real-time customer outcome-management strategy risking actual customer success, it is also throwing the repurchase cycle to chance. Whether customers are satisfied enough to buy more units, other products, upgrades, etc., is uncertain. And for those who do eventually repurchase, the time frame varies wildly. You will see just how wildly as you begin to gather the raw data for your DTR calculation.

Now back to our theoretical example above: As you can see, the customer ended up making two purchases (an initial purchase and one repurchase) in the period.

This would change considerably in a VAS world.

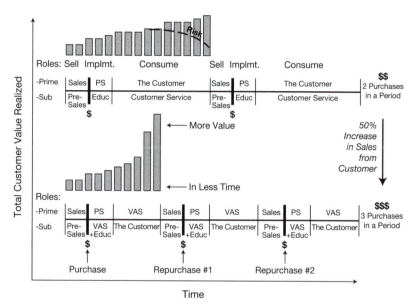

FIGURE 7.2 Accelerating DTR: The VAS Path to Customer Value.

To begin with, our company assumes an active management role in every phase of the process. First, we are using the VAS organization to backstop the professional services engagement, hopefully improving its margins on the project. At the same time, the VAS team is working with the customer to develop the initial "consumption plan" for the value realization phase of the adoption life cycle. The VAS teams, armed with usage data sent in by the products, then assume the PRIME role. They are monitoring the customer's actual utilization, looking for strengths and weaknesses. Those weaknesses are caught and addressed NOW— while the final outcome can still be influenced. They are reporting to the customer on their progress and working with the customer to facilitate successful value realization.

By inserting the VAS team into the PRIME role during the consumption (or value realization) phases of the life cycle, we do two important things:

- We increase the chance of customer success and decrease the risk of failure. (And that outcome is now much more under our control.)
- We accelerate the customer's time-to-value. (Again, we have a much higher degree of control.)

What is the net effect of these two results? A more successful customer, ready for more units or related products faster. It's simple: Implementing VAS will reduce DTR. Reducing DTR will increase customer base revenues. You will notice in the example above that the same customer has now made three purchases in the period, instead of the two purchases when the customer assumed the PRIME role during consumption. And it will also have a slew of other benefits to you like lower cost of sales and improved margins on repurchases, better customer referenceability … all the things we have touched on.

As we suggested, run some numbers for your company. Take a guess at your DTR and how much VAS could reduce it. Then calculate the impact on the Customer Base Revenue index and your corresponding revenues. The gross margins from that sales increase could put a real dent (or more!) in the cost of implementing VAS.

We think you will find that once you begin to monitor DTR, you will begin to accrue many of the company-wide benefits that you get from monitoring DSO. What makes DSO so useful is that it is a single number that reflects the health of several important business factors, including the quality of new business, product availability, ease of installation, and the effectiveness of accounting processes. Using DSO as a measure of financial health drives a lot of positive behavior in the sales, development, services, and accounting functions of your company.

DTR does much the same thing, plus some. It too is affected by all those factors, but it also takes into consideration the totality of business functions—both yours and the customer's—that impact the length of the repurchase cycle. Sales has to sell the right products, the products must be deployed effectively, the end users must use them, and the customer must get value. If customers aren't getting value out of products that they already own, they're not going to be as inclined to buy more or upgrade to the new version. But if everyone in your company is focused on driving down DTR, repurchase cycles accelerate, total customer base revenue grows, and you have more rapid growth from your most profitable source of revenue: existing customers. Each customer should be assigned a CBRI number at all times. When a customer falls below a certain level, they need your attention. No need for surveys, no months of analysis. It is like that individual customer's EKG. If it falters, you know it now and you take action to restore desired purchase behavior. It is a gas pedal for accelerating customer revenue. And, by the way, the customers will also be very satisfied and loyal, as your surveys will show.

It is time for tech companies to take ownership of this opportunity. Proactively managing DTR may be one of the biggest untapped levers to increasing corporate profits and shareholder value.

Remember: CBRI = ASP ÷ DTR. Measure people on it. Compensate people on it. We think you'll be amazed at the positive behavior it drives across your entire company. Your sales department will sell better. Services will provide faster implementation assistance. Development will build better usability and monitoring into the product. The big customer adoption problems will be identified and fixed. Individual customers will get real-time course correction. You will be—for the first time—in control of the consumption phase, and your entire team will begin to innovate around the activities needed to ensure that customer success becomes routine.

8 | Services: The Next "Big Thing" in Tech

"Moving to VAS would be too_____," you say?

Complicated? Expensive? Culturally difficult? Possibly. Will issues like VSOE make VAS even trickier to implement? Probably. Will it force major changes that might mess with cash-cow businesses like maintenance? Definitely.

VAS as a successful, scalable business model will not be easy to achieve. But we feel that ignoring the need for it is not an option for well-managed tech companies. The industry has a consumption gap that is real and growing. Tech products left and right are underachieving their potential. Customers have made it clear that they want to buy results, not technology. They have simply been outstripped by the complexity of the products or their own inability to change processes to take advantage of them. They can't keep up. Virtually no digital product can escape this trend. We have to move away from being product companies that invest only in product R&D. We must truly consider ourselves product AND services companies that invest in R&D across both disciplines, develop business strategy for both parts of the solution in concert, and measure investment returns in the broader context of account profitability.

There are other reasons why this is a unique time. As we have maintained throughout, the maintenance model is in serious trouble today. We need to replace that value proposition—to change those engines while the plane is still in the air. For enterprise technology companies in computing, medical, office solutions, and dozens of other categories, the time is now—before it is too late.

A third element that makes this a unique moment is the emergence of something called "service science." Just as the field of computer technology expanded and innovated exponentially following the creation of the "computer science" discipline in the 1960s, so too now will the field of services. Slowly but surely, again led by IBM, universities and academics around the world are working together to develop a curriculum known as SSME—Service Science Management and Engineering. Institutions like Cambridge University, Cal Berkeley, Carnegie Mellon, ISS in Germany and dozens of others are building curriculum and funding research in this rapidly growing field of science. They join early pioneers like the University of Leipzig, Rensselaer Polytechnic Institute, University of Maryland, Karlstad University and Arizona State University, who for more than 20 years have been working at proving that service is a science in every sense of the word. They are showing that the academic disciplines of engineering, operations management, economics, psychology, and over a dozen other established fields can be systematically applied to the development and optimization of services. This growing community includes not only academics but industry associations like TSIA, ACM, and IEEE. Also, many prestigious academic associations like INFORMS and POMS are developing service science communities. This is in addition to government and government-funded service research activities at places like the giant Fraunhofer Institute in Germany and the Industrial Technology Research Institute (ITRI) in Taiwan. Pulling this important but diverse community together to maximize its effectiveness is key. That's why TSIA co-founded, along with IBM and Oracle, the Service Research & Innovation Institute (SRII). SRII works with dozens of other corporate, academic, institutional and government partners to promote and coordinate global research in services. (For more on SRII visit www.thesrii.org.)

Why is this so significant? There are a lot of reasons, but one stands out.

I was chatting with the head of the Office of Innovation of CSC (Computer Sciences Corporation) two years ago. His CIO asked him a very fair but tough question: "How do I invest *strategically* in services?" It really made me think. I decided that there were two critical dimensions to this seemingly simple question. The first is around the issue of how. We know how to invest in software development, we know how to invest in hardware development, and we know how to invest in building Web sites. We know what skills we need, where to find them, and how to manage them. We know what tools and technologies to buy. We understand how to assess market opportunities for products, and how to collect design input from customers. We know how to estimate, schedule, and cost out a product development cycle. We even understand the best ways to pay for it. We are so confident in this ability that we allocate a fixed percentage of the total company budget to R&D. Great ideas or no ideas, product development gets funding.

But what about services? Can your company invest with the same degree of confidence in the development of services? Does your company understand the science of so-called "service systems"? Do you know how the field of psychology is to be applied to the service development process? Do you know how to test your services in a lab? Do you have people who are skilled experts in services development methodology? Do you know how the operations management academic discipline is adapting many proven manufacturing models to accommodate the services concept of value co-creation? Does your company even have a budget for services R&D?

If the answer to one or more of these questions is no, then you too may be struggling with the question of HOW to invest in services. You're not alone. In our recent landmark survey of service R&D practices, TSIA and the Service Research Innovation Institute (SRII) discovered that the industry overall is just getting started in this area.[1]

Service Research and Innovation (SR&I) Practices in Tech Companies

Parameter/Practice	%
Percentage of respondents with a dedicated SR&I group	31%
Percentage of services SR&I groups that are independent entities	3%
Percentage of respondents with a dedicated budget for SR&I	25%
Percentage of respondents that can measure the impact of investments in SR&I	39%

Source: TSIA-SRII

Only one in four tech companies has a budget for services R&D and less than 40% can measure the impact of the investments that they do make. The problem is deep. Services R&D does not come naturally to many companies. They report deficits in many basic areas needed to effectively innovate in services.

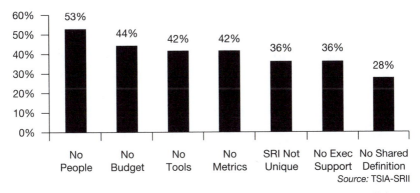

Source: TSIA-SRII

FIGURE 8.1 Which Challenges to Services Research and Innovation Exist Within Your Organization?

Most importantly, this survey evidences what we call the *service research and innovation gap*.

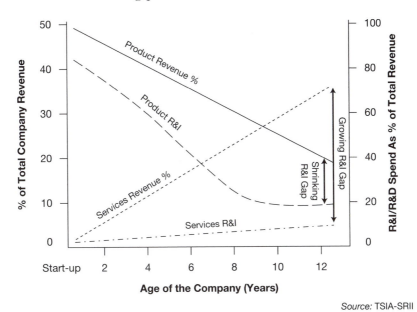

Source: TSIA-SRII

FIGURE 8.2 The Services Research and Innovation (R&I) Gap (Software Companies).

This chart illustrates the phenomenon using a fictitious software company. In its start-up years it is rightly dedicating all of its R&D dollars to building products. It eventually settles at investing 11% of annual revenues in product development. But as the company ages, the service businesses eventually grow and overtake product revenue in the total revenue mix. While service goes from zero to over 50% of revenue, it continues to receive virtually zero budget for research and innovation. More and more of the company's total financial performance is becoming dependent on an offering that is not a focus for the company's R&D efforts. While the possibility to innovate around services exists, the company exhibits the classic services R&I gap. For the first time, this growing gap is a real threat to the stock price of tech companies. There WILL be innovation in service over the next decade—real and noticeable innovation. Some companies will invest and reap the rewards while others fall behind. You can no longer count on growing your service revenue and profits tomorrow doing the same things you have done for the past 15 years.

So will service science play a critical role in closing this gap? Yes. Service science is on the verge of answering many of the tough questions we previously raised. It is safe to say that there are thousands of academic researchers working globally on these answers. Also, universities around the world are lining up to develop SSME* (service science, management, and engineering) curriculum to produce a labor pool for your company that is skilled and experienced in services R&D as well as all other key service roles. In the very near future, you will be able to invest with confidence in the services research and development function. We believe it will become inextricably joined at the hip with your product development activities. Soon you will know HOW to invest in services.

The second critical issue raised by the CSC CIO's question is this: "I know how to invest to improve operating efficiencies in services but how do I make *strategic* investments there?" Over the past decade we have done a wonderful job of applying technology and global labor arbitrage/sourcing to improve the economics of the customer service, education, and managed services functions. We know where to base our

---

* To learn more about Service Science and SSME, we suggest you visit the Service Research Innovation Institute at www.thesrii.org.

call centers, what processes to automate, and what tools to buy. The result has been substantial gains in both customer experience and financial outcomes. Twenty years ago, most customer support was available from 8 a.m. to 5 p.m., Monday through Friday. Today, service availability never sleeps. We move cases around the world as the sun moves; we have engineering collaborating across time zones—even across companies—to solve customer problems better, faster, and cheaper.

We have done similar things to make customer education more efficient. We now deliver computer-based training rather than classroom training as a way of dramatically expanding the number of users who can get trained. There are breakthroughs after breakthroughs in technology service efficiency—and most all of them stemmed from a desire to cut costs.

But how do we make *strategic* investments in services? How do we disrupt existing business models, enable new markets, or steal market share from the competition using our services? How do we build and retain defensible IP in services? Service science is also beginning to address these questions. Increasingly, services will be both the leading and the lagging edge of the technology adoption cycle. It will create new markets where none existed before. It will separate the commodity players from the value-creation players. In a moment, we will take a look at how.

SERVICES IN THE CLOUD

We could not finish the book without covering the hot topic of services in the cloud. Again, we are not talking about simply repackaging a product into a Web-based offering, billing by the use, and calling it a service. That's great! But it is still a product offering—just a repackaged one. We are talking about human-originated service. So does the highly touted model of cloud computing change anything in the outlook for the services that we are defining? Here is a view from Gartner:[2]

> *"The focus has moved up from the infrastructure implementations and onto the services that allow for access to the capabilities provided," said David Mitchell Smith, vice president and Gartner Fellow. "Although many companies will argue how the cloud services are implemented, the ultimate measure of success will be how the services are consumed and whether that leads to new business opportunities."*

Gartner predicts that the impact of cloud computing on IT vendors will be huge. Established vendors have a great presence in traditional software markets, and as new Web 2.0 and cloud business models evolve and expand outside of consumer markets, a great deal could change. "The vendors are at very different levels of maturity," said David Cearley, another vice president and Gartner Fellow. "The consumer-focused vendors are the most mature in delivering what Gartner calls a 'cloud/Web platform' from technology and community perspectives, but the business-focused vendors have rich business services and, at times, are very adept at selling business services."

What cloud computing represents is an opportunity to shift services up a notch in the value spectrum. All the principles of this book will apply—perhaps even more directly. The big advantage of SaaS and other Web-based computing models is that customers get to avoid a lot of the costs and pitfalls of the left side of the adoption cycle. There is no installation, integration is often made far easier, and maintenance is not the customer's problem. Those are real and formidable advantages. But it still does not address process change or any other keys to EUA. If we are smart, we will preserve the idea that customers still have to spend money on tech services, but the services are not for installation and implementation anymore. Instead, they should be targeted squarely at the value realization phase of the adoption life cycle.

A DOLLAR-IN OR A DOLLAR-OUT ORGANIZATION?

Let's face it, there are two different kinds of organizations in a company—the ones you put dollars into to grow profits and the ones you take dollars out of to grow profits. You invest money in R&D, sales, and marketing. You take money out of COGS, HR, and IT.

Where do services fall? For years, service has been a dollar-out organization. Beyond investments in efficiency improvements like technology, the goal was to optimize profits by reducing costs. You know what? That was probably the right thing to be doing.

But now, things are changing. If there is any one underlying point to this book it is this: Services is becoming a dollar-in organization. If you want to grow your value proposition, steal share from competitors, open new market segments, drive increases in revenues and profits for both product and services, then you need to restore the concept of investment to service. This transition from technology to results—and the rise of services that goes along with that—is happening today. Services can turn a great enabling technology into value for customers. This will be a cornerstone of industry growth in the coming years and decades.

WHERE IS IT HEADED?

If we only knew for sure exactly how things would unfold! But hey, what makes the tech business so interesting is that it is impossible to know what tomorrow holds. Maybe there will be a "next big thing" the size of the Internet that could change everything. Or maybe not. Maybe tech markets are really maturing and the rise of services is the natural evolution of the market—maybe services ARE the "next big thing." If we are right, if the rise continues, what does that mean to our core competencies as tech companies? Will they have to change? You bet.

As we have said often in this book, most of today's tech services don't actually create new value for the customer. That is up to the customers themselves. We do what we know we can repeat and scale, which is to help install and maintain our own technology. The fact that we have boxed in our services so tightly on things that we have done a million times means we really have not needed a science of services. But tomorrow we will. Here is why:

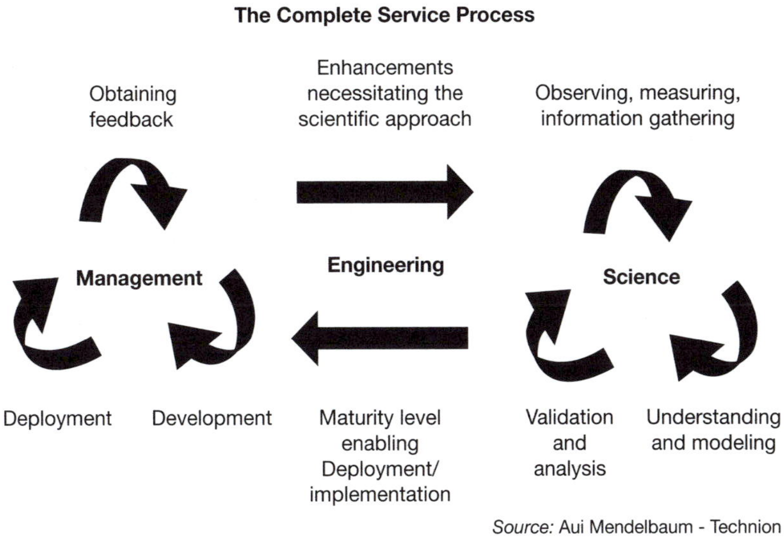

The Complete Service Process

Source: Aui Mendelbaum - Technion

FIGURE 8.3 The Complete Service Process.

A complete service process includes three fundamental competencies:

1. The science behind understanding an opportunity or problem and modeling a solution.
2. The engineering to make that model a usable, repeatable, and scalable service in a live customer environment.
3. Managing the process of implementing/delivering the actual service for the customer day to day.

Because we have stuck so closely to what we know well (our product) we have been able to spend most all our time perfecting the left side (Management) of the complete process. We didn't need science, we had experience. As long as we stuck to the product we were fine because it doesn't change that much from customer to customer. We even did some of the middle (Engineering) activities to build processes and tools that our service personnel and their customers could use to make the services more efficient, reliable, and affordable.

But what if the world of tomorrow is not just about delivering a working product but about actually helping customers get the potential value that the product could provide? What then? We as tech companies would have to obtain the ability to figure out how this one customer could take advantage of the product's value. We would have to know not only our product, but the customer and their industry. The number of services would—by necessity—rapidly proliferate. We would need a formal, repeatable process to "ideate" on a mass-customized basis (Science) across the multiple industries that buy our products.

That is what might scare you most about the world of VAS—the labor cost. Does HP need to staff its consumer customer service teams with photography experts? Does Oracle need bankers? Does GE Medical need doctors? And do all these industry experts have to be paired with a scientist who can apply "scientific rigor" to the process of uncovering customer opportunity or solving customer problems?

The answer is somewhat yes but mostly no. We will all need to move in the direction of increasing both the industry and application knowledge of our service teams. We will also need to invest in training to enable them to conduct competent discovery work on a customer's condition.

But mostly, it takes a village. This is another thing that IBM gets that many companies don't. A next generation of value-creation services is needed to pull more and more technology products (hardware and software) into the marketplace. We need to tackle new problems, not the same old ones. We need to fix traffic congestion, health care, and global warming. For the next few years, public-sector problems will offer huge growth opportunities for tech companies. Applying the complete service process to these problems will help solve them. Then it will enable software to be written, modified, and/or used to automate the solution. The software will need hardware. The hardware will need to be connected. The connections will need bandwidth, and so forth. But from here on out, services will be at both the front end and the back end of most companies' revenue growth.

Does IBM want to hire every known expert in every known field? No. What it wants is to build a global village around service science that allows it (and you!) to tackle new problems and opportunities for customers. It wants academics on the problems. It wants governments on the problems. It wants to grow the size of the IT market using services. It wants to be singularly qualified to be a resource integrator who can solve customers' most complex problems. It wants a global labor pool of people skilled in services and service science available to help either as employees, partners, or consultants. Just look at the following IBM slide.[3]

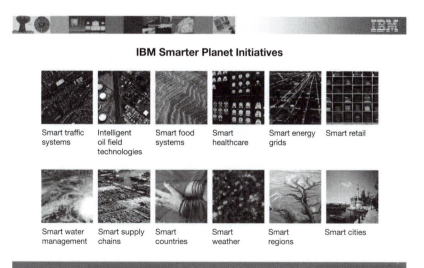

IBM Smarter Planet Initiatives

| Smart traffic systems | Intelligent oil field technologies | Smart food systems | Smart healthcare | Smart energy grids | Smart retail |

| Smart water management | Smart supply chains | Smart countries | Smart weather | Smart regions | Smart cities |

FIGURE 8.4

If you were the IT buyer for the city of Berlin and your local governmental leaders had an initiative around relieving traffic congestion, who would you turn to for IT solutions? Would it be a computer company that sold computers and some middleware or a computer company that had a specific practice in smart traffic systems? It is not about the technology anymore. More and more and cheaper and cheaper alternatives are out there for that. It is about the application of the technology to the unique business opportunities and problems of your customer.

Now maybe IBM can tackle an audacious goal like increasing the size of the IT spend in 12 major markets. What about you? How could services increase the size of your particular market(s)? How could services increase the EUA of your customers and help them survive the complexity avalanche? How could services help your products deliver new sources of revenue and profits? How will you develop your own village of service experts to make it all possible and affordable?

What does the rise of services mean to you?

"HOUSTON, WE HAVE A PROBLEM …"

We would like to leave you with an image—an image of our future, except the image is from the 1960s.

We would argue the first modern tech support organization was also the first VAS organization. It was in Houston, Texas, and it went by the name of NASA Mission Control. Remember all those dedicated men and women sitting at their stations monitoring every aspect of the mission, the spacecraft, and the astronauts?

What if NASA used the same approach as some tech companies? It would have sold the spacecraft to the astronauts, put it on the launch pad for them, strapped them in, got everything ready to go, counted down 5…4…3…2…1… and then left.

Thankfully NASA saw it differently. For Mission Control, the blast-off was the BEGINNING of the journey for their end-user customers (the astronauts). They helped with every aspect of the mission; they made sure the astronauts succeeded at each step. They proactively anticipated problems that their customers might encounter and worked out solutions ahead of time. If the customers struggled with a task, they knew it and they helped. They understood the outcomes the customers wanted and they measured them diligently. They made sure the mission was as successful and valuable as possible. THEY NEVER LEFT THE CUS-TOMER UNTIL THE MISSION WAS FULLY ACCOMPLISHED. And once that mission was over, they started developing plans for the next one.

That is value added service and that is the image you should keep in your head—those people with short haircuts glued to their stations and obsessed with customer success. Our future versions of VAS will be more efficient, more profitable, and far simpler to execute. But the mission remains the same: Using technology services to help your customer not just survive the complexity avalanche, but to get the maximum value from your great products.

9 | Not Your Father's Industry Association

LIKE MANY BUSINESS PEOPLE, WHEN SOMEONE MENTIONED THE TERM "industry association" to me I thought of two things: government lobbying and cocktail parties. I was wrong.

Industry associations are one of the most powerful (but too often poorly executed) business models out there.

When a tech company joins the Technology Services Industry Association (TSIA) we don't just demand its money. We demand data, practice information, and intellectual engagement at many levels. Each of the nearly 400 global service organizations in TSIA provides an average of 200 data points per year to our in-depth collection of industry benchmarks and studies. We produce well over 100 reports and analysis per year on trends in service strategy, operations, technologies, and performance levels.

When we built TSIA we didn't hire people with experience at nonprofits. Our staff comes from companies like Apple, HP, IBM, Oracle, and Sun. Our researchers don't come from academia; they come from tech analyst firms like IDC, Forrester, and Aberdeen. When we did see an opportunity to foster breakthroughs in service science by bringing the global academic community together with industry and government, we co-founded the nonprofit Service Research & Innovation Institute (SRII).

AN N OF MANY

Why is the industry association such a potentially powerful model? Simply, it is the power of the community. Let's face it, industry associations were the original B2B communities. No one company has all the right answers. No one company has all the best practices. No one company has all the innovations. And no one company can see all of what's coming next. Each company is an N of one. By engaging in TSIA we are able to pull together an N of many. We put a thought from this company together with a thought from three hundered others. Pretty soon things become clear. We collect real and meaningful data sets. Want to know what the average gross margin is for professional services organizations inside software companies? Want to know what new technologies are being successfully adopted to increase customer self-service success rates using rich media? Want to know how to apply VSOE to project-based services? Want to know how companies are compensating their sales forces on services? Want to know the best practices in defending against maintenance discounts? We can help. We have even put all the best practices together in a single Organizational Development Program so that customer service organizations can go directly to world-class without the trial and error of figuring it out themselves.

We think the association model is unique. We can gather data that no research company can and make sense of it quickly. While we don't do consulting per se, we partner with leading companies like PRTM that do. We are often asked to bring the facts, and steer the findings on consulting engagements for our members.

And because our fact-based model is funded by so many companies, the ROI is easy to see.

THE SWITZERLAND OF THE INDUSTRY

We have found that companies almost always have gaps in their understanding about the state of the service art and science. We have also found that every company has a contribution to make to the state of the service art and science. What they needed was a safe place to deposit their insights and pose their questions. And they needed some smart people who could look at what everyone contributed and make some sense of it all. At TSIA we provide those things to member companies large and small. We don't invite the press, we don't invite the financial analysts, and we don't make private data public. We are a safe, warm place to admit your company's not perfect and to learn from your peers. It is powerful, and we (TSIA and our members) are excited about it. If your service organization is not involved, we strongly encourage you to check us out. There is a reason why virtually every major technology company in IT and a growing number of other tech markets are already active members of TSIA.

And yes, we do have cocktail parties.

Endnotes

Chapter 1

1. Interview with Don Norman, author of *Design of Everyday Things*. August 2008.

2. TSIA/Neochange/Sand Hill Group. Q3 2009. "IT Success Survey."

3. NPD Group Inc. 2009. *Mobile Phone Usage Report* press release. p. 1.

4. WebUser.com. 2008. "Millions of Gadgets Unused in Britain." British Telecom Survey.

5. Accenture. 2008. "Big Trouble with 'No Trouble Found' Returns–Confronting the High Cost of Customer Returns." p. 4.

6. Brombacher, A.C., Sander, P.C., Sonnemans, P.J.M., and Rouvroye, J.L. 2005. Managing product reliability in business processes "under pressure." *Reliability Engineering & System Safety* 88.

7. Neil, D. February 2009. "50 Worst Cars of All Time," Time.com.

8. DeSisto, R. 2006. Quote to CRM News.

9. Cisco Systems and Momentum Research Group. 2009.

10. MO Council's Forum to Advance the Mobile Experience. 2007. "The Global Mobile Mindset Audit" Executive Summary. p. 6.

11. TSIA/Neochange/Sand Hill Group. Q3 2009. "IT Success Survey."

Chapter 2

1. CIO Insight. "Top CIO Priorities for 2009."

2. Bailey, G., and Hagen Wenzek, H. 2005. "Irresistible." IBM Press. p. 2.

3. Books by Geoffrey Moore include *Crossing the Chasm, Inside the Tornado,* and *Dealing with Darwin.*

4. TSIA/Neochange/Sand Hill Group. Q3 2009. "IT Success Survey."

5. TSIA/Neochange/Sand Hill Group. Q3 2009. "IT Success Survey."

6. Chris Dowse interview with Dean Lane, author of *CIO Wisdom.* 2006. Neochange.com.

7. TSIA/Neochange/Sand Hill Group. Q3 2009. "IT Success Survey."

9. Really Simple Systems and AMI Partners Survey. November 2007. www.reallysimplesystems.com/release.asp?id=8.

10. Crowell, B.A. "A Structured Approach to Physician Adoption of Technology." The Advisory Board Company. p. 1.; "Clinical Advisory Board Essay: The Case for Reform." Washington DC: Clinical Advisory Board.

11. May, S. February 2002. "Patient Safety Drives Refinement of Systems." *Healthcare Informatics.*

12. Kohn, L.T., Corrigan, J., and Donaldson, M.S., eds. 1999. *To Err is Human: Building a Safer Health System.* National Academy Press: Washington.

13. Knoa Inc. and Reachforce study of 1,000 global IT executives and business stakeholders. October 2008.

14. TSIA/Neochange/Sand Hill Group. Q3 2009. "IT Success Survey."

15. Garbani, J. 2007. "Appliance-Based End User Experience Monitoring." Q2 2007.

16. ECPweb, Macrovision, SoftSummit, CELUG, and EDA Consortium. November 2008. "Key Trends in Software Pricing and Licensing."

17. ServiceSource. 2008.

18. Duggan, J. August 2008. Gartner Research Presentation: "Application Portfolio Management and Enterprise Architecture—Agility Through Understanding."

19. Garbani, J. 2007. "Appliance-Based End User Experience Monitoring." Q2 2007.

20. Gartner. August 2008. "Dataquest Insight: IT Markets Remain Resilient in 2008 and Will Grow Moderately in the Next Three to Five Years."

21. Gartner. August 2008. "Dataquest Insight: IT Markets Remain Resilient in 2008 and Will Grow Moderately in the Next Three to Five Years."

22. Public financial reports by Oracle and SAP.

23. TSIA Research. Q2 2009. TSIA "Services 50."

24. TSIA Research. Q1 2008. TSIA "Services 50."

25. Consumer Electronics Association and Services Revenue. 2007. "Beyond Delivery and Installation: Premium Services Consumers Want." p. 20.

26. TSIA Research and Decision Strategies International. 2006. "Navigating Uncertainty—Future Scenarios for Technology Services."

27. Associated Press. September 5, 2008. "Microsoft Gurus to Take On Apple's Geniuses."

28. Dunn, B., Best Buy CEO, quoted in the company's Q1 F08 earning call.

29. TSIA member company. Used with permission.

30. TSIA member company. Used with permission—EMC.

31. Womack, J., and Jones, D. 2005. "Lean Consumption." *Free Press*. p. 9–17.

32. Neochange/Chasm Institute. 2008. "Enterprise Software at the Tipping Point."

Chapter 3

1. TSIA Research. 2009. SSPA Benchmark Database.

2. TSIA Research. 2009. SSPA Benchmark Database.

3. TSIA Research. 2009. SSPA Benchmark Database.

4. TSIA Research. 2009. SSPA Benchmark Database.

5. Ricketts, J.A. 2008. "Reaching the Goal: How Managers Improve a Services Business Using Goldratt's Theory of Constraints."

Chapter 4

1. Dowse, C. 2008. "Increasing Software Provider Profits through Effective Usage." Neochange.

2. SAP Web site at www.sap.com.

3. TSIA Member Companies. Public information or company approved.

Chapter 5

1. Wikipedia. 2009. http://en.wikipedia.org/wiki/Software_metering.

2. IBM Web site. 2009. www.redbooks.ibm.com.

3. Glushko, R.J., and Tabas, L. June 15, 2007. "Bridging the 'Front Stage' and 'Back Stage' in Service System Design." School of Information. Paper 2007-013. http://repositories.cdlib.org/ischool/2007-013.

4. TSIA Research. 2009. TPSA Benchmark Database.

5. CA Web site. 2009. www.ca.com/us/about/content.aspx?cid=190253.

6. Mills, S. 2008. "Future of Business." IBM.

7. Wang, R. July 2009. "An Enterprise Software Licensee's Bill of Rights, V2." Forrester.

8. TSIA/Neochange/Sand Hill Group. Q3 2009. "IT Success Survey."

Chapter 6

1. TSIA Research. Q2 2009. TSIA "Services 50."

2. International Labour Organization, U.S Department of Labor, IBM.

3. TSIA Research. Q2 2009. TSIA "Services 50."

4. IBM—Company approved. 2009.

5. Quote attributed to author Peter Drucker. http://wiki.answers. com/Q/What_gets_measured_gets_managed.

Chapter 8

1. TSIA Research and Service Research Innovation Institute (SRII). 2009. "The Services Innovation Gap." p. 23.

2. Smith, D.M. June 26, 2008. "Gartner Says Cloud Computing Will Be as Influential as E-business." Gartner.

3. IBM—Company approved. 2009.

Index